Technological Change in the Modern Economy

T0260823

Technological Change in the Modern Economy

Basic Topics and New Developments

Paul Beije

Associate Professor of Management of Innovation, Erasmus University Rotterdam, The Netherlands

Edward Elgar

Cheltenham, UK • Northampton, MA, USA

Published by
Edward Elgar Publishing Limited
Glensanda House
Montpellier Parade
Cheltenham
Glos GL50 1UA
UK

Edward Elgar Publishing, Inc.
6 Market Street
Northampton
Massachusetts 01060
USA

A catalogue record for this book
is available from the British Library

Library of Congress Cataloguing in Publication Data
Beije, P. R. (Paul R.). 1952–
    Technological change in the modern economy : basic topics and new developments / Paul Beije.
       1. Technological innovations—Economic aspects. I. Title.
HC79.T4B43 1998
   338'.064—dc21
                                                 98-38242
                                                      CIP

ISBN 1 85898 331 2

Printed and bound in Great Britain by
Biddles Ltd, Guildford and King's Lynn

Contents

List of Figures and Tables

FIGURES

TABLES

Preface

This book is the result of over five years of experience with teaching an elective on 'the economics of innovation' at Erasmus University Rotterdam, initially at the Economics Department and the last three years at the School of Management. The selection of the various topics and the way in which they are treated has been gradually developed over the years and by now it can be said that the book gives a fairly complete overview of the economics of technical change. The text provides an introduction for undergraduate students, for managers and for scholars interested in the topic of innovation and technical change.

Apart from the fact that not many introductory texts are available on this topic, the book more specifically offers two perspectives that are not common in current textbooks. First, it deals with the various topics from an empirical point of view and theoretical explanations are treated in depth only when they give support to empirical studies. Other theoretical literature will be dealt with briefly but the reader is provided with a list of further readings at the back of each chapter. Second, the book presents data from European countries while most of the textbooks are US-oriented. Further, it examines theoretical studies from both anglosaxon and continental European countries. It also includes some Japanese work.

Although the main focus is on economic problems of innovation, other aspects are dealt with too. Concepts from the management of innovation literature are included in the discussions when they clarify things from the economic approaches to a topic, or when they bring in aspects that are rather neglected by economists. Elements from other social sciences are also included in some of the discussions. In general, an institutional approach to economics is taken that facilitates the integration of elements from various social sciences. The institutional approach is taken rather loosely, however. The reader is introduced to the main institutions around innovation and technological change. Within this framework of 'technological institutions' several economic theories are positioned, such as the neoclassical theory of technological change, the evolutionary theory, and transaction cost economics from the new institutional economics.

The topics can be divided into two parts. The first six chapters contain the basic issues from 'the economics of innovation'. Chapter 1 gives an introduction to innovation and technological change. The other chapters in this part discuss topics well known in the economics literature, such as the appropriability problem and the use of patents. Chapter 1 is a must for readers without any prior knowledge of innovation and technological change. Others can skip through the first chapter just to get acquainted with my terminology. Chapters

7 to 11 deal with new topics, such as technological cooperation, spillover effects of research, and learning. Although some of the chapters from this second part could be read on their own, in many cases the knowledge contained in the first part is a prerequisite for a better understanding of these new topics. Various insights from the basic topics on innovation presented earlier in the book are used to explain technological cooperation, spillover effects, and so on.

At the end of the journey through 'the land of innovation' many people deserve credit for the knowledge I tried to expose in this book. The many students from the economics department, from the management school, and from foreign universities, who followed my course, contributed directly by providing feedback in class. Indirectly my colleagues and the secretaries at the Department of Institutional and Industrial Economics (EOV) and at the Department of Management of Technology and Innovation (MTI) contributed to the book, because they gave me time and space to fulfill the task and helped me to become the economist I am. All the graphs were produced by Raoul Manten who also did all other layout jobs, before the people from Edward Elgar Publishing succeeded in making the book look like it is now. Edward, thank you for trusting me to do the job, particularly when it took more time than I myself had thought. Dymphna Evans and Francine O'Sullivan, thank you for given me the feeling that you were always there to solve my problems immediately. Fiona Peacock and Emma Gribbon also deserve credit. All the people mentioned until now played an essential part in the production of this book, but the book could not have been written without the support of my relatives and close friends who have supported and encouraged me throughout. I thank them all, without exception.

Paul Beije
Rotterdam

1. What is innovation?

1 INTRODUCTION

The aim of this introductory chapter is to give the reader a first understanding of what innovation is. It forms the basis for the other chapters. I first introduce the basic terminology, largely following the economics of innovation literature. Thereafter I turn to practice. Section 3 looks at the persons and organizations that create innovations and at the kinds of activity they undertake. The next section introduces some theoretical elements about innovation activity. Two models are presented which make a systematization of various interrelated innovation activities. The last part, Section 5, briefly discusses the main motives to carry out innovation activities.

2 BASIC TERMINOLOGY

Innovations are new things applied in the business of producing, distributing, and consuming products or services. They can take many different forms, such as new management procedures, new financial services, new distribution facilities, new products, and so on. In fact, modern societies change continuously. The focus of this book is on innovations in technology. *Technological innovations* are defined as new products (called product innovations) and new machines or equipment (processes innovations) or improvements of existing products and processes, which have been established on the basis of some technological change created by the innovator, and which are commercially exploited.

All of us are confronted with new products. For example, in the market for personal computers new PCs arrive every year. The most important change compared to the former type of PC is the capacity of the memory chip and the speed with which it operates. But several other elements may have been renewed, such as the monitor. Most of us do not perceive new processes, but they are equally important from an economic point of view. New machines may become available from which (new) PCs are made more cheaply or with higher quality.

The innovator can be a person or an organization. The technological change he brings forward is basically new knowledge.[1] By definition a process innovation is created and used 'in-house' by the innovator. An example is a technological improvement made by a company of its own machines at the production floor.

1

Product innovations are by definition all new or improved products that are realized by the innovator, but used by others, usually the customers of those new or improved products.[2] Innovations differ with respect to the nature and extent of technological change they embody. The degree of newness of the product or process relies on the extent of technological change established by the innovator. In many cases innovations consist of some new element created by the innovator and some or many technological elements and aspects which have already been applied before (often by others).[3] An example of the latter is the use of an existing component (for example an integrated circuit) in a new product.

A technological innovation can be more or less complex. Complexity of a product is sometimes defined as the number of its components and the inter-relations between the components. According to Mansfield (1968, 72) the improvement of an existing product will imply a considerable degree of new-ness when the original product was already complex, because 'If there is con-siderable interdependence, a change in one part requires the redesign of other parts'.

The distinction between *radical innovations* and *incremental innovations* can be based on the 'degree of newness' in the same way. An incremental inno-vation aims at improving one or a limited number of parts. A radical innova-tion is concerned with a completely new design of the product. New parts nor-mally have to be designed and the relations between the parts may be new too.

More classifications of technological innovations exist which are based on various criteria of newness, but they can also be based on the economic impact of the innovations. We come back to several of them in the following chapters.

Inventions are technological achievements that have not been commercially exploited yet. They often include a large part of the technological newness

Figure 1.1 Inventions and innovations

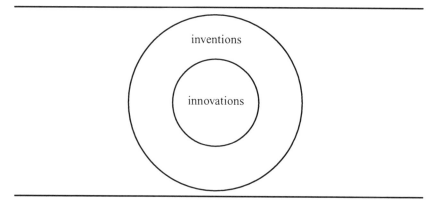

underlying new products, new processes, and their improvements, but also technological achievements without any further commercial success. It is assumed here that all innovations embody an invention of the innovator and subsequently that an invention completely determines the degree of newness of the innovation.[4] This does, however, not imply that all inventions are embodied in an innovation.[5] Figure 1.1 gives an abstract picture of this.

Economists usually put the emphasis on innovation, not on invention. Schumpeter (1952), one of the first economists who gave full credit to the role of technological change in economics, sees the role of the innovator as crucial for the economy, not the role of the inventor. The innovator applies 'neue Kombinationen' which in many cases may be based on an invention that already existed and was applied in another industry before. The difference between innovator and inventor is formulated as follows (Schumpeter, 1952, 129):

> Economic leadership in particular must hence be distinguished from 'invention'. As long as they are not carried into practice, inventions are commercially irrelevant. And to carry improvement into effect is a task entirely different from the invention of it, and a task, moreover, requiring entirely different kinds of attitudes. Although entrepreneurs of course may be inventors just as they may be capitalists, they are inventors not by nature of their function but by coincidence and vice versa. Besides, the innovations which it is the function of the entrepreneurs to carry out need not necessarily be any inventions at all.[6]

The description by Jewkes et al. (1969, 30) puts emphasis on some other aspects of the difference between invention and innovation:

> The one [invention] is a beginning, without which the other is of no avail; is in the nature of things largely unpredictable; arises more immediately out of technical preoccupation and the combination of that with the sense of artistry, craftsmanship and making things fit which is characteristic of much invention. The other will be informed at each step by economic calculation, each successive step becoming more deliberate and specifically designed to a defined end.

The *innovation process* is the activity or set of related activities that leads to a single innovation or a number of interdepending innovations. An innovation project can be seen as a particular set of activities aiming at the realization of a single innovation. As each innovation contains one or more inventions, the innovation process can be seen as a mixture of two quite distinct kinds of activity, as the foregoing quote of Jewkes et al. made clear. Large firms usually undertake many innovation projects simultaneously.

For the time being we define *Research and Development (R&D)* as the set of related activities undertaken to create technological innovations. Today many firms have an R&D department in which these innovation activities are concentrated. We have already seen that a difference exists between invention and innovation. Consequently, there may be actors playing the role of inventor and of innovator. In practice invention and innovation are not homogeneous activities. Different kinds of activity are needed to invent or innovate and most of these activities need coordination. Consequently, many different types of innovation activity exist. Usually the different types of innovation activity are not undertaken by a single unit of the innovating firm. Moreover, some activities take place outside the firm. For the time being we assume that all innovation activities are undertaken by the R&D department of the firm and concentrate on the differences between the R&D activities 'basic research', 'applied research', and 'development'.

Basic research is the activity aiming at the enhancement of knowledge without any (direct) application in new products or processes. It includes most *scientific research* and *generic technology* projects. In both cases the outcome of this activity is new knowledge. There is a difference between the two. Most scientific research is not aiming at the creation of knowledge with which to make new products and processes, and in most instances it is carried out by universities and other public research institutes. Generic research aims at the creation of a new technological base with which (many) technological innovations can be strived for. When private firms undertake basic research, it is usually generic technology they want to create. Hence, a difference exists between science and technology. According to Polanyi (1969) the purposes, methods, and results of pure science and technology differ. Pure science is directed towards understanding, and technology is directed towards use. To emphasize this difference one usually speaks of scientific discoveries to indicate newly created knowledge in science, and inventions to describe new knowledge related to new technology. In the latter case the new knowledge may be seen as incorporated in an invention.[7]

Applied research is directed at creating new technology with the purpose to apply it in new products or processes. It is the activity which, although based on scientific principles on many occasions, aims at commercial exploitation of the inventions which as such are the outcome of this kind of research. *Development* is the term used to describe the activities undertaken to commercialize a technological invention. Applied research can be said to result in 'prototypes' of new products or processes, while development is the activity within the R&D department undertaken to refine and test the prototype before commercialization can be realized. An example of a refinement is the use of another material than the original prototype contained. This material does not alter the functions the prototype can perform, but it changes its duration or makes it otherwise better adjusted to the conditions under which it is supposed to be used in practice (by customers). The relations among science, basic research, applied research and development are discussed later in this chapter. Various other activities may relate to innovation, such as engineering, design, and software development. They are explained later in the book.

Not all R&D projects have an equal 'amount' of basic or fundamental research, applied research, and development. In general it can be said that the greater the kind of technological change aimed for in an innovation project the more basic research will be involved. According to Roussel et al. (1991, 15) *incremental R&D* contains very little research and mostly development activity. The goal is small advances in technology, based on the clever application of existing knowledge in science and engineering. 'A typical example of incremental R&D is work on reducing manufacturing costs. Most manufacturing processes can be improved by a continuous series of small but important advances: energy conservation, computer-guided process control, better metallurgy for lower maintenance costs.' *Radical R&D* contains a large amount of both 'R' and 'D'. The foundation of existing scientific and engineering knowledge is insufficient to arrive at the desired practical result. 'The work undertakes the discovery of new knowledge with the explicit goal of applying that knowledge to a useful purpose' (ibid., 15). *Fundamental R&D* contains merely 'R' and is a 'scientific/technological reach into the unknown' (ibid., 16).[8]

This classification can, in our opinion, be clarified by the list of objectives of the various R&D activities shown in Table 1.1.

Table 1.1 R&D activities and their objectives

Type of innovation activity	Objective
Fundamental research:	knowledge with no direct commercial application
Applied research:	new technology to be used in new products and processes
Development:	commercial application of new technology in new products and processes

One could say that basic or fundamental R&D is quite different from development activity and even more different from many other activities within the firm. During the last century most activities within the firm have been planned, monitored, and measured, but the more innovation contains 'R' the more difficult it becomes to fit this activity within the control logic of the (large) firm. Roussel et al. (1991, xi) describe the situation as it developed until the 1970s and 1980s as:

Western business executives have been indoctrinated in the concept of management based on measurement. Measurements of activity (for example, sales or units
produced) serve as surrogates for measurement of productivity. Cost accounting
and control systems have been extended into practically every corner of the enterprise. The research and development function, however, has characteristically
resisted this pressure for short-term measurable results, because the results most of
the time cannot be seen to be counted. Other functions in the business resent the
R&D resistance to being held accountable on comparable terms.

In Chapter 3 we come back to the question whether and how R&D can be
managed. It turns out that the more 'D' a project contains the more planning
and monitoring devices can be used, although they still differ substantially
from procedures and methods in production, inventory and other routine activities.

A basic concept in economics not immediately related to the creation of
an innovation is *imitation*. Imitation is the activity aiming at direct replica of
an existing innovation or at the creation of a similar new product or process.
The latter are based on what the imitator knows about the innovation and aim
at commercial application in the field where the original innovation is used
too. In many occasions the imitator is a competitor in the market of an innovating firm. Put in other words: imitation is about the use by others of the knowledge created by the innovator. The more imitation occurs the less the profits
for the innovator are. Imitation can therefore be said to be concerned with the
economic effect of innovation.

The same holds for the concept of *diffusion*. The diffusion of an innovation is the widespread use of the original innovation in the economy. Diffusion
may be purposefully strived for by the innovator, as in the case of a new product
he sells on the market, but it may also be involuntary, for example when a new
machine is imitated by many other firms. The more the innovation is used in
the economy by people other than the innovator the greater is the diffusion.

3 MAIN ACTORS AND ACTIVITIES IN PRACTICE

After the basic vocabulary has been established we can now turn to look at
which individuals, private companies, government organizations, and other
institutions are undertaking R&D and how these activities differ.

The actors

Many innovations are the work of individuals or small teams. They may be
working on their own, in small firms, or in large corporations. Jewkes et al.

(1969, 11) have added ten case descriptions to their 1958 first edition of the book, which are exemplary of the various sources of innovation/invention[9]:

> Float Glass and Semi-Synthetic Penicillin are instances of successful invention and development by firms of considerable size. Air-Cushioned Vehicles, Computers, Photo-Typesetting and the Wankel engine are cases where independent inventors made important advances at the earlier stages and both large and small companies made contributions in the later development. With Oxygen Steel-making and Chlordane, Aldrin and Dieldrin the discovery and development can be attributed to comparatively small firms. The histories of the treatment for Rhesus Haemolytic Disease and the Moulton Bicycle are largely records of the achievements of independent workers.

Today the majority of innovations are created by private companies. In a considerable number of innovations individuals within firms still play a crucial role, but teamwork tends to become more important. Individuals operating outside firms regularly come up with new technological findings (inventions), but it is rare to see them develop their prototype into full commercial production. Therefore they are called individual inventors, not individual innovators. Small firms are also largely inventors and not innovators, with the exception of so-called science-based firms. The reason behind the fact that innovation by individuals and small firms are exceptions has to do with the high cost of development, as we will show later in this section.

Jewkes et al. (1969) carried out a large and well-known study about the sources of 70 important innovations in the twentieth century.[10] From the 70 cases more than one-half (38) can be ranked as individual inventions, carried out by a single person or by a team of researchers independent of private firms; 24 of the inventions have their origin in research laboratories of large and small manufacturing companies; the remainder are difficult to classify (ibid., 73–7). In some industries, however, where large industrial research laboratories are quite common, the importance of individual inventors seems to have declined.

Jewkes et al. (ibid., 81) rightly state that it is not obvious what exactly one means by individual inventor:

> In one extreme sense every inventor is an individual inventor and every invention is an individual invention: since all human minds function independently, a new idea must arise in one brain . . . The adjective 'individual' must obviously apply to the conditions under which the inventor does his work: whether he is self-employed or works as an employee under contract for some other individual or institution; whether he is free to do what he wishes or is under agreement to think and work within prescribed lines; whether he works in a large team or a small; whether, with-

in the team, he is one of many subjects under the control of others or is the head of a group following his instructions and providing his ancillary services.

In Chapter 3 we discuss the different approaches of Usher to the (individual) act of invention. There it will become clear that an individual inventor may be following his or her own direction, but he/she is certainly influenced by and communicates with other specialists in the field.[11]

Jewkes et al. (1969, 141–2) conclude that research in private firms (systematically stimulating the stream of inventions by individuals) has been increasing, but is no complete substitute for the efforts of independent individuals, for three reasons:

1. From 1900 to 1960 R&D in large corporations has not been responsible for the greater part of important inventions.
2. These firms continue to rely heavily on outside sources of original thinking.
3. Large firms may themselves be centres of resistance to change.

More recent studies (Macdonald, 1986; Sirilli, 1987; Amesse et al., 1991) support the general conclusions of Jewkes et al.: the share of individual inventors in (patented) inventions is declining and the minority of them are really innovators (carrying the invention through all the stages needed to commercially apply the invention).

Development activity is completely different in kind from research, as will be discussed further in Chapter 3. The importance of independent individuals is less in this phase of the innovation process and seems to have been declining in this century. According to Jewkes (ibid., 161) 10 of the 70 important

Table 1.2 Development costs of some specific innovations

Innovation	Period	Estimated cost
Burton petroleum cracking process	1909–13	$ 92,000
Houdry petroleum cracking process	1925–36	$ 11,000,000
Du Pont Orlon fibre	1941–47	$ 5,000,000
Ampex video-Tape recording	1951–55	$ 1,000,000
GM diesel-electric locomotive	1930–34	$ 4,000,000
DC8 aircraft	1955–59	$ 112,000,000
Beecham's semi-synthetic penicillin	1947–57	£ 2,500,000
ASCC computer	1937–44	$ 400,000
Atlas missile	n.a.	$ 3,000,000,000

Source: derived from Jewkes et al. (1969, 214–16)

inventions have been developed further without enormous costs by individuals and smaller firms.[12]

The major part of the innovation activities undertaken by individuals and private companies are development, and applied research, or activities not officially classified as R&D, such as design, and engineering (for quantitative data see Section 3). Universities and public/private research institutes, although largely concerned with basic and applied research and not development, regularly come up with inventions, and occasionally with innovations. The role of public and private institutions is much greater than the number of innovations they realize suggests. We will see later that the knowledge created by such institutions may be an important information source to the innovation process of private companies.

After our first impression of which persons and organizations are creating innovations we now look more closely at the innovation process, at the size and nature of the activities which are undertaken to create innovations.

The activities

Today the largest part of innovative activities is registered as R&D, and although individual inventors still play an important role for some types of innovation, in terms of money spent they are not important any longer. This has to do with the fact that most R&D expenditure is devoted to 'development' and individual inventors lack the resources to undertake this activity. In addition, the costs of development have been rising the last one hundred years (Jewkes, ibid., 155). Of course, as said before, individuals may play an important role in the R&D carried out within companies.

At the end of the nineteenth century some large firms established laboratories (Mowery and Rosenberg, 1989) and during the next century the number of laboratories rose sharply. Table 1.3 presents figures for the USA. Today most large firms carry out innovation activities in R&D departments; there may be a central laboratory (which was the case in the aforementioned establishments in the US), but also several divisional R&D departments. Figure 1.3 gives an example. As we will see later, many small firms do still not have an R&D department.

Various authors have described this shift of innovation activities from individuals to firms in terms of the professionalization of R&D (see, for example, Freeman, 1982 and Mowery and Rosenberg, 1989). In fact innovation and inventive activities of autonomously operating individuals are not registered and it is the official registration of R&D activity in the (larger) firm which forms the base for many economic analyses of innovation. Freeman (1982, 10–11) associates the professionalization of industrial R&D with three main changes that took place:

Table 1.3 Laboratory foundations within manufacturing industry (mfg.), 1899–1946

		Prior to 1899	1899–1908	1909–1918	1919–1928	1929–1936	1937–1946	Total (% of total mfg.)
Food & beverages	number	11	20	32	50	48	40	201
	% of row	5.5	10.0	15.9	24.9	23.9	19.0	(8.7)
Tobacco products	number	0	0	1	2	3	1	7
	% of row	0.0	0.0	14.3	28.6	42.9	14.3	(0.3)
Textile products	number	3	4	11	16	28	17	79
	% of row	3.8	5.1	13.9	20.3	35.4	21.5	(3.4)
Apparel	number	0	0	1	1	0	2	4
	% of row	0.0	0.0	25.0	25.0	0.0	50.0	(0.2)
Lumber products	number	0	1	1	2	5	5	14
	% of row	0.0	7.1	7.1	14.3	85.7	35.7	(0.6)
Furniture	number	0	0	0	2	1	1	4
	% of row	0.0	0.0	0.0	50.0	25.0	25.0	(0.2)
Paper	number	4	6	15	38	26	13	102
	% of row	3.9	5.9	14.7	37.8	25.5	12.7	(4.4)
Publishing	number	0	0	0	2	3	1	6
	% of row	0.0	0.0	0.0	33.3	50.0	16.7	(0.2)
Chemicals	number	40	56	88	178	146	107	615
	% of row	6.5	9.1	14.3	28.9	23.7	17.4	(26.7)

Petroleum	number	5	3	15	25	31	10	89
	% of row	5.6	3.4	16.9	28.1	34.8	11.2	(3.9)
Rubber products	number	2	2	16	19	13	5	57
	% of row	3.5	3.5	28.1	33.8	22.8	8.8	(2.5)
Leather products	number	3	0	4	9	3	1	20
	% of row	15.0	0.0	20.0	45.0	15.0	5.0	(0.9)
Stone, clay glass	number	5	12	24	54	39	12	146
	% of row	3.4	8.2	16.4	37.0	26.7	8.2	(6.3)
Primary metals	number	9	19	30	42	29	14	143
	% of row	6.3	13.8	21.0	29.4	20.3	9.8	(6.2)
Fabricated metals	number	0	17	24	53	37	28	159
	% of row	0.0	10.7	15.1	33.3	23.3	17.6	(6.9)
Nonelectrical machinery	number	6	14	49	65	63	30	227
	% of row	2.6	6.2	21.6	28.6	27.8	13.2	(9.9)
Electrical machinery	number	12	18	28	53	64	44	219
	% of row	5.5	8.2	12.8	24.2	29.2	20.1	(9.5)
Transportation equipment	number	4	4	12	16	10	20	66
	% of row	6.1	6.1	18.2	24.2	15.2	30.3	(2.9)
Instruments	number	6	4	17	23	32	36	118
	% of row	5.1	3.4	14.4	19.5	27.1	30.5	(5.1)
All manufacturing	**number**	**112**	**182**	**317**	**660**	**590**	**388**	**2,303**
	% of row	**4.9**	**7.9**	**16.1**	**28.7**	**25.6**	**16.8**	**(100.0)**

Source: Mowery and Rosenburg (1989, 62)

Figure 1.2 Technology transfer from central R&D to product division within the firm

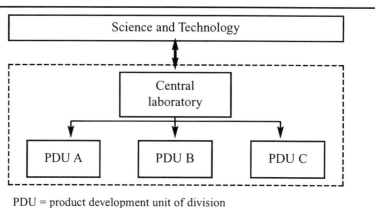

PDU = product development unit of division

1. The increasingly scientific character of technology;
2. The growing complexity of technology;
3. The general trend towards division of labour.

This led to a relative decline of public R&D activity (including scientific research) and a strong growth of R&D activity performed by private firms. It also made the interactions between private firms' R&D units and public research units important, as we will see later.

Most of the R&D expenditure of private firms is devoted to applied research and development. However, dividing R&D into basic research, applied research, and development activity is difficult, because of a lack of data, but also because the borderlines are difficult to define precisely. Most firms that spend money on R&D are undertaking development activity, fewer of them are occupied with applied research, and only a minority is involved with basic research.[13] Figure 1.3 depicts DuPont's annual investment in basic and applied research and development for nylon during the period 1928–48. It stands out clearly that expenditure on development is considerably larger than expenditure on research. Two other things are revealed by the investment curves in Figure 1.3, which have not been discussed in this chapter yet. First, development continues after nylon has been introduced in the market at a commercial scale. Second, investment in plant and facilities is larger than total R&D expenditure.

In addition to basic and applied research and development, various authors from the management literature come up with a fourth category of innovation activity, namely 'technical service'. Ramsey (1986, 2) gives the following description:

Technical service involves research that is used almost exclusively to support present sales effort. An example of technical service research is the modification of a current product for a new customer. Technical service research is normally located within the operating division rather than at the central R&D location, and it is normally a part of the operating, rather than the research division.

Here we have an example of product improvements that are realized outside the R&D department and without any activity included in the definition of R&D. In the same way process improvements may be originating from an engineering department. Hence, while product and process improvements may be called incremental innovations, they are partially realized by people who do not undertake R&D.

The pattern of R&D expenditure in the nylon case is not atypical for the innovation process in general. Table 1.4 gives some figures about the division

Figure 1.3   Investment for the development and production of nylon

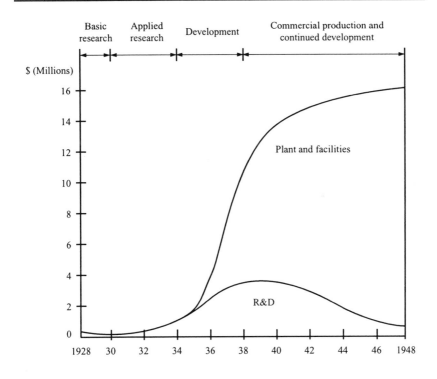

Source: Scherer (1984, 4)

Table 1.4 Distribution of gross expenditure on R&D by source of funds (percentages)

	Industry			Government			Other nat. sources			Abroad		
	1981	1985	1991	1981	1985	1991	1981	1985	1991	1981	1985	1991
United States	48.8	50.0	52.6	49.3	48.3	44.9	1.9	1.7	2.5	..	..	..
Canada	41.3	40.4	41.3	50.0	47.6	44.0	4.9	4.2	4.7	3.9	7.8	10.0
North America	48.4	49.5	52.1	49.3	48.3	44.8	2.0	1.8	2.6	..	..	..
Japan (adj.) [1990]	67.7	74.0	77.4	24.9	19.1	16.4	7.3	6.8	6.1	0.1	0.1	0.1
Belgium [1990]	..	66.5	70.4	..	31.6	27.6	..	0.8	0.7	..	1.1	1.3
Denmark	42.5	48.9	51.4	53.5	46.5	39.7	2.0	2.6	4.6	2.1	2.0	4.4
France	40.9	41.4	42.5	53.4	52.9	48.8	0.6	0.8	0.7	5.0	4.8	8.0
Germany	57.9	61.8	60.5	40.7	36.7	36.5	0.4	0.3	0.5	1.0	1.2	2.5
Greece	21.4	25.6	21.7	78.6	74.4	57.7	..	..	0.7	..	..	19.9
Ireland	37.7	45.7	59.4	56.5	46.1	28.2	1.1	1.5	2.1	4.8	6.6	10.3
Italy	50.1	44.6	47.8	47.2	51.7	46.6	..	..	..	2.7	3.6	5.7
Netherlands	46.3	51.7	51.2	47.2	44.2	44.9	1.3	1.5	1.9	5.2	2.6	2.0
Portugal [1980, 1984, 1990]	22.6	30.8	27.0	66.8	62.1	61.8	4.7	4.7	6.5	1.9	2.4	4.6
Spain	42.8	47.2	48.1	56.0	47.7	45.7	0.1	0.2	0.6	1.1	4.8	5.6
United Kingdom	42.0	46.6	49.3	48.1	42.8	35.5	3.0	2.7	3.7	6.9	7.9	11.5
EC	48.3	50.9	51.7	46.9	44.2	41.1	1.1	1.0	1.2	3.7	3.9	5.9
Austria	50.2	49.1	50.3	46.9	48.1	46.5	0.4	0.3	0.3	2.5	2.5	2.9
Finland	54.5	..	56.3	43.4	..	40.9	1.1	..	1.5	1.0	..	1.3
Iceland	5.7	24.1	24.5	85.6	64.3	69.7	5.0	8.7	1.7	4.3	2.8	4.1
Norway	40.1	51.6	44.5	57.2	45.3	49.5	1.4	1.0	1.3	1.4	2.1	4.6
Sweden	54.9	60.9	60.5	42.3	36.4	35.3	1.4	1.5	2.7	1.5	1.2	1.5
Switzerland [1989]	75.1	..	74.5	24.9	..	22.6	..	..	1.3	..	..	1.5
Turkey	..	..	28.5	..	..	70.1	..	..	1.3	..	..	0.2
Nordic Countries	50.6	57.4	55.4	46.5	39.5	39.5	1.5	1.6	2.6	1.5	1.4	2.4
Australia [1984, 1990]	20.2	28.0	40.3	72.8	68.5	54.9	2.1	2.4	3.6	1.0	1.1	1.3
New Zealand	..	..	32.9	..	..	65.2	..	..	..	..	..	1.9
Total OECD	**51.2**	**54.0**	**56.9**	**45.0**	**42.2**	**38.1**	**2.4**	**2.4**	**2.8**	**1.3**	**1.4**	**2.2**
OECD Median	**42.5**	**47.2**	**49.3**	**50.0**	**47.6**	**44.9**	**1.4**	**1.5**	**1.7**	**2.1**	**2.5**	**3.5**

Source: OECD (1994).

of R&D into the categories 'public' and 'private' in OECD countries. Note that no distinction is made between basic and applied research, and development. It is known from various other studies that most basic research is performed by or financed by government (see Chapter 6).

It can be assumed that indeed most development and a large part of applied research are undertaken by private business. Where the output of basic research is in most cases technological knowledge that is used somewhere and sometime in applied research and development to create innovations, the outcome of most applied research and all of development is new products and processes. Not all kinds of firm create innovations in equal amount. The size of the firm may count, and also the industry in which the firm operates.

Table 1.5 provides figures about the percentage of private firms with R&D expenditure in different European countries by size class. Two things need to be emphasized here. First, more firms spend money on R&D in higher size classes than in the smaller size classes. Second, it is possible that small firms which do perform R&D spend relatively more than large firms.[14]

Today an important question in relation to economic policy is whether a country needs small firms that undertake R&D. Individual inventors undertake applied research, but in many occasions merely 'practical thinking' to come up

Table 1.5 Firms with R&D in different size classes in various European countries[1]

Size classes		NL.	N.	DK.	IE.	A.	G.
10–19	R&D(o)	9.4	11.3	NA	20.0	10.7	27.0
	R&D(p)	56.1	54.2	NA	64.0	NA	85.0
20–49	R&D(o)	19.0	22.0	25.8	21.0	24.9	28.0
	R&D(p)	64.7	53.1	56.1	64.0	NA	69.0
50–99	R&D(o)	38.2	39.5	41.6	27.0	47.7	24.0
	R&D(p)	71.6	51.6	57.7	76.0	NA	83.0
100–199	R&D(o)	48.3	45.0	52.9	45.0	59.6	43.0
	R&D(p)	79.0	75.5	74.4	75.0	NA	86.0
200–499	R&D(o)	59.5	56.6	33.3	52.0	71.9	55.0
	R&D(p)	80.5	92.7	68.7	91.0	NA	92.0
≥ 500	R&D(o)	66.4	69.0	59.0	52.0	90.4	78.0
	R&D(p)	85.0	86.3	91.1	96.0	NA	97.0

R&D(o): percentage of firms in each size class that undertake R&D occasionally
R&D(p): percentage of firms in each size class that undertake R&D permanently
NA: not available
[1] NL.=Netherlands, N.=Norway, DK.=Denmark, IE.=Ireland, A.=Austria, G.=Germany

Source: Kleinknecht (1996).

with inventions, especially for new or improved products. This holds for small firms too. Small firms (up to approximately 100 employees) used to realize quite a number of inventions and important innovations, as Jewkes et al. (1969) and others showed. Acs and Audretsch (1988) state that almost half the number of innovations are contributed by firms smaller than 500 workers. These innovations are on average as significant as the innovations of large firms. One needs to realize, however, that most firms (>90 percent) are small and that a majority of them do not undertake any R&D at all, or their (R&)D is devoted to getting incremental innovations. These are realized without any substantial research and are the result of development, design, and engineering activity. Today small firms in new technological fields like microelectronics and biotechnology are exceptions. They perform applied and even basic research. One can state that if private firms are undertaking basic research at

Table 1.6 R&D intensities by industry in 1994 in selected countries

Industry	US	Japan	Denmark	France	Germany	Italy	NL.	Spain	Sweden	UK
Total manufacturing	8,0	7,3	4,2	6,8	6,2	2,7	4,9	1,4	10,4	5,4
Food group	1,2	1,7	1,5	0,9	0,4	0,4	2,2	0,3	1,5	0,9
Textile group	0,6	1,6	0,1	0,9	1,4	0,1	0,9	0,5	1,3	0,3
Wood group	0,4	1,1	0,3	0,6	0,5	0,1	0,1	0,2	0,4	0,5
Paper group	1,3	0,8	0,2	0,3	0,3	0,1	0,3	0,2	1,9	0,2
Industrialchemicals	8,0	12,7	3,4	10,3	9,1	3,1	9,9	0,9	6,0	7,3
Pharmaceuticals	23,7	20,3	26,4	27,5	15,8	14,2	15,9	3,8	39,0	33,3
Petroleum refining	9,3	5,9	0,0	1,5	0,6	3,0	2,0	0,7	2,2	4,2
Rubber and plastics	3,1	5,1	2,1	3,7	1,9	1,1	1,7	0,7	5,1	0,9
Non-metalic mineral	2,1	4,2	0,6	2,1	1,4	0,2	0,6	0,5	1,6	1,1
Basic metals	1,6	4,6	2,4	2,9	1,0	0,8	5,5	0,9	3,1	1,3
Fabricated metals	1,3	1,6	0,3	1,2	1,4	0,6	1,0	0,7	1,9	1,6
Non-electrical mach.	4,0	8,3	6,7	6,9	6,8	2,0	2,3	2,3	9,0	4,5
Off. and comp. equipm.	49,5	24,4	7,2	11,3	17,6	12,3	44,5	4,6	65,5	5,9
Electrical machinery	5,9	12,1	3,6	4,6	11,6	3,4	1,8	8,2	8,6	
Radio, TV equipm.	15,0	15,4	17,6	34,2	14,8	25,3	7,6	13,0	67,4	13,9
Shipbuilding	0,0	1,5	4,5	4,5	3,4	13,0	0,5	13,5	3,6	2,3
Motor vehicles	16,5	10,0	0,0	12,6	13,6	10,9	17,0	2,1	19,1	9,7
Aircraft	36,1	16,9	0,0	37,6	43,4	39,1	12,5	39,8	52,0	22,1
Other transp. equipm.	3,8	4,5	7,5	8,1	9,1	3,4	0,0	3,4	9,9	3,6
Scientific instruments	21,0	18,3	14,2	4,4	4,2	1,3	3,9	6,8	24,9	3,7
Other manufacturing	3,7	1,5	12,4	1,0	1,5	0,3	3,0	0,9	1,3	0,5

Business enterprise R&D expenditure as a percentage of value added

Source: OECD, 1998, Annex Table 4.13

all, they are either big or very small, and, in the latter case, mostly operating in so-called high-tech industries.

Industries differ considerably in the degree to which R&D is undertaken. Some industries, like furniture or leather, largely consist of small firms and this may explain the relatively low degree of innovation. But Acs and Audretsch (1988) conclude that in some industries small firms realize the majority of innovations while they are of minor importance in other industries. A relatively low level of R&D expenditure is also explained by the *technological opportunities* of the industry, and by other factors, such as the degree of competition, government regulation, nature and growth of demand. For the time being we only show the differences between industries' R&D efforts and postpone an explanation of these differences. Table 1.6 presents R&D intensities according to industries in European countries. R&D *intensity* is defined as the ratio of the average R&D expenditure of the firms in an industry and the average volume of sales of these firms. R&D intensity is used instead of R&D expenditure to make a correction for size differences between firms or industries. Without this correction it is expected that large industries will perform more R&D than small industries.

We assumed earlier that firms spend more money on development that on applied and basic research. This could be due to the fact that a considerable number of innovation projects of a typical firm do not include any research activity, or most projects do include some research, but the expenditure for this activity is on average less than on development.

4   A SIMPLE MODEL OF INNOVATION ACTIVITY

Now we know about the basic concepts, and about the main things that are going on in practice, a first attempt is made to more systematically deal with some major innovation activities. This 'modelling' of innovation activity mainly concerns firms and not individuals or not-for-profit organizations.

We begin with the so-called stepwise model that reflects a particular (and simple) view on what a typical innovation project may look like. Then we extend this model by making a distinction between the innovation project as such, the marketing activity related to the project, and the set up of a production line for the innovation. We conclude this section with a brief introduction to the co-ordination of multiple projects in a firm.

The stepwise model of innovation activities

Each innovation project can be thought of as a number of more or less sequential steps. These steps cover the major activities needed to realize an innovation. Most of the time the model is used to describe a product innovation. It begins with a stage of idea generation from which the main outlines of a possibly commercially viable product development project follow. The basic idea includes an overall picture of what the new product is technically able to do and what the product 'architecture' would look like. In this model ideas sometimes follow from scientific development and basic research activities. In the next stage of (applied) research various possible alternatives of details of product architecture and functional performance are explored and a preferred prototype of the new product is the result. Earlier we have called this the invention. In the next stages this prototype is made ready for the market.[15] First the prototype is extensively tested under circumstances similar to those under which customers of the new product will operate. This stage is called development. It may include the organization and testing of the production line required for the manufacturing of the new product on a commercial scale. The last stage concerns the marketing of the new product. The main characteristic of the model is the ordered way in which the whole innovation process is undertaken: each stage results in something from which it is decided whether or not to start the next stage. Figure 1.4 reflects this structure. Moreover it is assumed that a particular team undertakes one stages and gives the results to the team that is supposed to do the next stage. The model is said to reflect the technology push vision on innovation: from scientific and technological developments ideas for concrete commercial development projects are derived and the new product merely follows from those technological developments. By contrast demand pull innovations are undertaken because knowledge about market developments generate ideas about new products. An example is the improvement of a board computer of a truck on the basis of specific problems truck drivers currently have in using the computer.

The stepwise model focuses on the 'technological newness' of the prototype. The successive steps of further development of the prototype and other things to bring the new product to the market are seen as necessary but not crucial. It is implied that a project results in a single innovation, namely the invention or prototype, which is developed into commercial application. The example of the invention of nylon below shows that such a clear distinction

Figure 1.4 The stepwise model of innovation

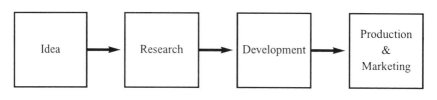

between invention and further development of the invention (without any new inventions) is not always possible in practice.

The example of nylon   (Source: Jewkes et al., 1969, 30–31)

In 1935, after seven years' work of varying fortune with many disappointments, work which might have led anywhere or nowhere, W.H. Carothers, in the laboratory of the du Pont Company, produced the first nylon fibre and du Pont undertook to translate it into a marketable product. By 1939 large-scale production of nylon hosiery had commenced. Thus, in a matter of four years of development, du Pont had reached its appointed goal. Estimates put the total cost of the early stages of research and development at over $1 million; at that time 230 technical experts were engaged in the work. What precisely was involved in the development undertaken after Carothers's initial discovery?

First, it was necessary to find ways of producing on a large scale the intermediate constituents of nylon which, up tot that time, had been made only on a small scale. The two important materials were adipic acid and hexamethylenediamine. Adipic acid had been manufactured in Germany for some time but there had been no commercial exploitation of it in the United States. The German processes were not readily adaptable to the plants of du Pont and it became imperative to develop a new catalytic technique for the purpose. Hexamethylenediamine posed even greater difficulties; it was merely a laboratory curiosity and had never been manufactured on a commercial scale before. Success here required the discovery of new catalysts and the proper handling of heat transfer problems. Next, a great deal of work had to be done at the stage where the materials react to form the long chain of molecules of the nylon polymer. The first polymers were made in glass equipment in Carothers's laboratory, but glass equipment was completely unsuitable for commercial manufacture and metal equipment had to be designed. Methods of controlling the degree of polymerization had to be evolved, since the failure to stop the reaction at precisely the right time resulted in the production of different and far less useful polymers than nylon. The technologists had to learn how to make one batch of the product exactly like another.

At the next stage of manufacture the flakes of the polymer had to be melted and some means found to transfer the molten mass to the spinning machines. Only pumping gave the filaments adequate uniformity, but unfortunately there were no existing pumps suitable for the task. A new type of pump was required embodying new alloys capable of withstanding the heat of the molten polymer. At the next stage of spinning, the machinery had to be specially designed for the task, since nylon could not be spun in the same manner as cotton, wool, viscose or cellulose acetate. The winding and the cold drawing processes also confronted the developers with problems which were novel and for which specially designed machines were required. Thus at each one of these stages – the mass production of what had formerly been made only on a small scale, the maintenance of unusual degrees of purity, the flexible controlling of the chemical process and the devising of mechanical aids for handling materials with novel properties – the developers were confronted by one hurdle after another. It was only when the process reached the stage of knitting and weaving that existing and familiar techniques could be called in to help. But at every stage workers knew what they were looking for, and, with varying degrees of certainty, they knew it could be found.

An extended model

The stepwise model discussed in the previous section may have been exemplary for many innovation projects in the past, but today innovation projects tend to be seen in a more integrative way. Development of the prototype, re-organization of the production line, of distribution channels, of relationships with suppliers of core components, and so on are seen as one system. The focus is on integration of simultaneously undertaken activities, and not on sequential steps of rather different kinds of innovation activity. We discuss

Figure 1.5 A more complex model of innovation in the firm

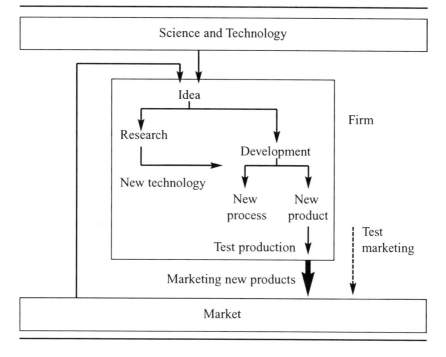

these matters more in detail in Chapter 3. Figure 1.5 reflects the more systemic approach of the innovation process.

One of the implications of the extended model is that an innovation is no longer the end product of a final stage of activity. It is possible that innovations occur at various places in the system (such as the system in Figure 1.5). The example of nylon has clearly shown that quite a number of (incremental) innovations in process technology and equipment had been necessary before the prototype nylon could be produced on a large scale. Pavitt (1982) must have had such cases in mind when he stated that innovations not only occur at the end of the applied research stage (the so-called invention stage), but also during testing and pilot production.

It is important to note that the precise meaning of innovation may change with the model of the innovation process used. In the stepwise model the innovation is the new product at the end of the chain of activity. In the extended model the aim may still be to market a new product, but innovations may arise in engineering, testing procedures of the prototype, and other activities carried out during the innovation project as well.

The coordination of multiple projects

All but very small innovating firms have a number of innovation projects most of the time. The R&D expenditure of a typical firm i is therefore the result of the costs of the N innovation projects it carries out per time period:

$$R\&D_{it} = C(p_{1t}) + C(p_{2t}) + \ldots + C(p_{Nt}) \qquad (1.1)$$

Each project needs to be planned and managed separately, but together they also need to be coordinated in several dimensions. An example is the planning of a time schedule in which it has been laid out when prototype testing is carried out for each project, given that they use the same test facility that has limited capacity.

In Chapters 2 and 3 we will see that there are several types of projects (such as incremental innovation projects and radical innovation projects mentioned earlier in this chapter) which, among other things, may burden the overall R&D budget differently.

5   WHY IS ONE UNDERTAKING INNOVATION ACTIVITIES?

As we have seen, the majority of innovations stem from private firms. Why do such firms undertake R&D investments, which are costly and whose results are unknown by definition? A large part of the R&D of private firms is devoted to projects aiming at commercial development of new products and processes. The subsequent question is therefore: why are firms developing new products and processes? In practice a variety of motives of different managers, research teams, and others may exist. The overriding motive for innovation, according to the economics of innovation literature, however, is the strengthening of the firm's position in its market(s). This motive may be coming from the inherent desire to be the best in a technological sense, or it may be forced by the innovation performance of competitors who are threatening the market share and profit of the firm. Improving one's competitive position and other motives are dealt with in various other chapters.

6 CONCLUSIONS AND OUTLOOK

The most general description of innovation is 'the process to undertake a change in one or more of the many aspects of the production, distribution, and consumption of economic goods'. Schumpeter (1968) has made a classification of innovation that is more practical, consisting of (1) new products and processes, (2) new distribution methods, (3) ways to penetrate new markets, and (4) the use of new management practices and organizational structures. It contains all basic categories of ways in which the entrepreneur can earn money by undertaking new activities. In this chapter we have taken a much more limited view. Only new products and processes based on a change in technology were examined. This limitation notwithstanding, the current state of knowledge on innovation offers a picture of increasing complexity.

The role of inventors has been gradually taken over by R&D departments of firms. Today innovative activities of large firms consist of many teams and various R&D units. The coordination of these different innovation activities and with production and marketing becomes more and more complex. We have also seen that innovation activities consist of more than R&D. The economic literature tends to focus on R&D as the only kind of activity. For the time being we look at R&D as the only set of activities for innovation, but we noticed that within this set different kind of activities are undertaken by different actors (basic and applied research, development).

When the R&D unit of the firm is seen as the producer of innovations several comparisons between firms and industries can be made. R&D intensity is a relative measure that can be used at the level of the firm and the industry and, as we will see later, at the level of the whole country. The R&D intensities of industries showed that some industries spend more money on R&D than other industries. The figures for different size classes made clear that the percentage of small firms undertaking R&D is less than the percentage of large firms.

In Chapter 2 the relation between innovation activity and economic growth is examined. Again the firm's innovation processes are simplified and reflected by the R&D intensity. The analysis is extended to relationships between firms in the creation and use of innovations. Chapter 3 offers a more detailed description and analysis of the innovation process. After reading that chapter more will be known about what lies 'behind R&D'.

NOTES

1. 'Technological innovation' and 'technical innovation' are often used interchangeable. I follow Freeman (1982, 4) who makes a distinction between technology as a body of knowledge about techniques (technological innovation) and technical innovations (which are the techniques themselves) that embody this knowledgde.
2. The limitations of our approach can best be illustrated by referring to the kinds of innovation Schumpeter (1952, 100) sees as 'neue Kombinationen' realized by the entrepreneur: (1). the production of new or improved products; (2). the adoption of a new method of production in an industry; (3). the entrance of a new sales market for an industry; (4). the exploitation of a source of raw materials or parts, new to the industry; (5). establishing a new form of organization in the industry, for example a cartel.
3. The inventor borrows much from others and it is difficult to find out precisely how much has been the result of the inventor's creativity and problem solving capacity and how much has been invented by others.
4. Although this may be true on many occasions, in the further development of a prototype, that is, the invention, additional innovations may be realized. For example it may turn out during the testing of a new prototype that the current material of some component does not qualify for the required 'zero defect' production and in a subsequent research effort it is discovered that another material is better suited.
5. It may be possible in practice that an invention exists of a new combination of existing technologies that works immediately, without any further adjustment or tests. It can thus be turned into an innovation quite easily if these technologies have all been commercialized before. It is possible therefore that innovations exist which do not embody an invention. An example is a 'double deck train' in which no new technology is involved and merely the idea of an additional 'level' has been added.
6. In the original German book (Schumpeter, 1968, 88–89) the text is somewhat different, but the main erea of difference between inventor and entrepreneur/innovator remains: Die Funktion des Erfinders oder überhaupt Technikers und die des Unternehmers fallen nicht zusammen. Der Unternehmer kann auch Erfinder sein und umgekehrt, aber grundsätzlich nur zufälligerweise. Der Unternehmer als solcher ist nicht geistiger Schöpfer der neuen Kombinationen, der Erfinder als solcher weder Unternehmer noch Führer anderer Art. Sowohl was sie tun, ist verschieden, als auch die Eignung zu dem, was sie tun – 'Verhalten' und 'Typus'.
7. The distinction between science and technology suggests that scientific knowledge may later be transformed into technological knowledge, but not the other way arround. However, in practice technological innovations also affect scientific developments. Another complication is that next to science basic research may be focused on understanding of technical artifacts, such as computers.
8. The distinction between fundamental, radical, and incremental R&D is quite similar to the one between basic research, applied research, and development, if one assumes that the research in fundamental R&D is first of all 'basic' in nature, and that in radical R&D 'radical'.
9. The main part of the book is about invention and not the development into full commercial application.
10. With regard to 'important' they say: 'The choice of what might be regarded as 'important' inventions has (...) had to be largely arbitrary. Every case chosen has been a commercial suc-

cess, shows promise of becoming so, or is obviously a discovery of great public value. Inventive virtuosity has not been accepted for its own sake (ibid., 67).

11. This is also stated by Jewkes et al. (1969, Chapter 5).
12. It is not said how small these firms were. The inventions thus successfully developed were: air conditioning, automatic transmissions, Bakelite, 'Cellophane' tape, electric precipitation, magnetic recording, power steering, quick freezing, shell moulding and the synthetic light polariser (Jewkes et al., 1969, 161).
13. Various studies have made estimations of the shares of fundamental research, applied research, and development, ranging from 5-20% for fundamental research, 20-40% for applied research, and 40-60% for development.
14. It is no surprise that large firms spend more on R&D than small firms. We introduce R&D-intensity in Table 1.3 in which a correction is made for the size of the firm.
15. We will see later that this 'technology-push' view of a product innovation project can be transformed into a more balanced view by including a 'business analysis' after the stage of idea generation which accounts for the market, that is, demand side.

REFERENCES

Amesse, F., C. Denranleau, H. Etemad, Y. Fortier, and L. Seguin-Dulude, (1991), The individual inventor and the role of entrepreneurship: A survey of the Canadian experience, *Research Policy* 20, 13–27

Acs, Zoltan J. and David B. Audretsch (1988), Innovation in large and small firms: an empirical analysis, *The American Economic Review* 78, 4, 678–690

Freeman, Ch. (1982), *The Economics of Industrial Innovation*, second edition, London, Francis Pinter

Jewkes, J., D. Sawer and R. Stillerman (1969), *The Sources of Invention*, London, Macmillan

Kleinknecht, A. (1996), Innovatie, imitatie en R&D samenwerking: Nederland vergeleken met vijf andere landen, in: F.J.M. Zwetsloot, *De Macht voor Wetenschappelijk Onderzoek*, Utrecht, Lemma, 201–227

Mansfield, E. (1968), *The Economics of Technological Change*, London, Longmans

Macdonald, S. (1986), The distinctive research of the individual inventor, *Research Policy* 15, 199–210

Mowery, D.C. and N. Rosenberg (1989), *Technology and the Pursuit of Economic Growth*, Cambridge (MA), Cambridge University Press

OECD (1994), *Science and Technology Policy; Review and Outlook*, Paris

OECD (1998), *Science, Technology and Industry Outlook*, Paris

Pavitt, K. (1982), R&D, patenting and innovative activities, *Research Policy* 11, 33–51

Polanyi, M. (1969), *Knowing and Being: Essays*, London, Routledge & K. Paul

Ramsey, J.E. (1986), *Research and Development Project Selection Criteria*, revised edition, Ann Arbor, UMI Research Press

Roussel, Ph.A., K.N. Saad and T.J. Erickson (1991), *Third Generation R&D: Managing the Link to Corporate Strategy*, Boston, Harvard Business School Press

Scherer, F.M. (1984), *Innovation and Growth: Schumpeterian Perspectives*, MIT Press, Cambridge (MA)

Schumpeter, J.A. (1952), *Theorie der Wirtschaftlichen Entwicklung: eine Untersuchung über Unternehmergewinn, Kapital, Kredit, Zins und dem Konjunkturzyklus*, Munich, Duncker & Humblot

Schumpeter, J.A. (1968), *The Theory of Economic Development*, 8th printing, Cambridge (MA), Harvard University Press

Sirilli, G. (1987), Patents and inventors: An empirical study, *Research Policy* 16, 157–74

2. Innovation and economic growth

1 INTRODUCTION

After our presentation in the first chapter of main types of actors and the size and nature of innovation activities, we want to give insight into the economic effects of innovation. A direct economic effect of innovation is the employment needed for innovation activities, which first of all is the employment in R&D. The economic importance of technology, however, lies more in its indirect (employment) effects than in the immediate innovation activities themselves. For example, employment in R&D is less important than employment in production in industries that survive because of the innovations the firms create through R&D.

The effects of innovations on economic growth can be assessed at the level of the firms that create them, and at the level of the economy as a whole. The main effect at the firm level is that innovations enable a firm to improve its competitive position and consequently increase real output and/or net profits. A firm, as we have seen, tries to calculate the benefits and costs of an innovation project; benefits relate to the firm's position compared to its competitors.

At the level of the entire economy the benefits of innovation for society matter and for a single project these may differ from the benefits of the innovating firm. We discuss the benefits for society. Innovation has two main effects on the economy. First, it provides consumers with new products and thus with a changing, and usually larger, set of choices. Second it makes available new machines, equipment, and material with which goods are produced. The new machines and equipment may improve the quality of the products (new and old!) or lower the production cost for the customers.[1] Note here that many new products in the economy, because of diffusion, may thus improve the efficiency of the firms who buy (or imitate) them. According to Scherer and Ross (1990) the distribution of industrial inventions by intended users is as shown in Table 2.1[2]:

Inventions in materials for use in other industries are not only new materials, such as glass fiber, but also new parts, subsystems, and systems created by suppliers. Examples are new integrated circuits and disk drives to be used by the PC manufacturers and a newly designed dashboard to be used by the automobile manufacturers.

The problem in more details

Economics has been much more involved with causes and effects of innovation than with the activities related to innovation themselves. In Chapter 1

Table 2.1 The distribution of innovations according to use

Production process for internal company use	26.2%
Capital goods for use by other industries	44.8%
Materials for use by other industries	21.6%
Consumer goods only	7.4%

Source: Scherer and Ross (1990, 616)

examples were given of how the R&D intensity of a firm or industry is used in statistics on innovation. The use of this 'abstraction' of the R&D process itself is largely due to the desire to compare firms or industries quantitatively with respect to innovation. Figure 2.1 gives a basic framework of cause and effect of innovation, while the process is approximated by the use of R&D (intensity).

From the section on the history of technology in economics it will follow that the long-term effects of technical change on the growth of GDP by country and later by industry have been the focus of economists. By now economists agree fairly well on the importance of these effects for whole nations. More complex, and giving rise to more disputes, is the question of how the process of innovation at the micro level leads to a country's long-run economic growth. One needs to understand how the technological results of firms are obtained and how the interactions between various firms and industries lead to macroeconomic growth. A complication here is that industries differ with

Figure 2.1 Growth of the firm

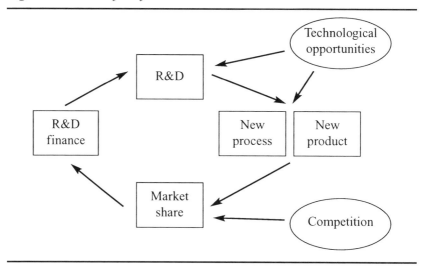

respect to the nature of the innovation process, the main sources of innovation, and their effects on the economy. We lack a generally accepted model which incorporates the most important strategic considerations of innovating firms and the impact of their R&D on profits and other indicators of firm performance, let alone that we are capable of linking such microeconomic processes and effects to macroeconomic effects of innovation. The current state of affairs is that we obtain positive statistical correlations between R&D and economic growth at the industry and macro level without a fully accepted microeconomic theory of innovation.[3] It is important to note that the macroeconomic effect of technology not only follows from a country's performance in innovation activities, but also from its efficiency to apply new technology. The literature at this point is dominated by studies in which R&D expenditure and productivity growth are linked in so-called production functions. These are estimated at the firm level, at the level of industries, and for countries as a whole. One of the disadvantages of production functions, when seen from the perspective of innovation, is that they are difficult to interpret as the reflection of innovative behaviour by firms.

At the micro level other theories of innovation have been constructed and tested. These theories try to incorporate variables about strategic decisions of innovation and R&D. An example is the microeconomic model of innovation that considers private returns to R&D as the main variable to understand decisions made by firms or in whole industries to invest in R&D. These estimations of private costs and benefits of innovation projects can be thought of as the core element underlying Figure 2.1.

Such microeconomic theories, however, do not explain macroeconomic growth. It is difficult enough to explain the relationship between innovation and growth at the firm and industry level. But this relationship is precisely the main focus of many economists involved with innovation: they want to measure the effect of an increase of R&D expenditure on the number and quality of innovations and subsequently on profits and other commercial performance indicators of firms or whole industries. The micro models try to explain why individual firms invest in R&D. They try to determine which factors influence the R&D investment decision and the profits gained from innovation.

The importance of innovation at the micro level lies in the increase of profits or market share by individual firms. The importance of innovation at the macro level goes further. Technical change is initiated by innovations, but its effects ultimately depend on the extent of diffusion or imitation of these innovations and the extent to which one innovation may stimulate a whole chain of subsequent innovations. An important role in this respect is played by the relationships between the creator and users of new technology. In this chapter we limit ourselves with respect to this aspect to the presentation of empirical evidence about technology flows between innovation-producing industries and innovation-using industries. In Chapter 3 we examine these flows further, because relations between firms in different industries are becoming an important topic of management and organization of innovation and technology of the individual firm. It shows that the nature of the relationship between creators and users may strongly influence innovativeness as well as the efficiency of use of new technology. In Chapter 4 the same issue is taken up, but then in terms of private and social costs and benefits of innovations; this is the approach usually taken by economists.

Innovation not only has a significant impact on economic growth, but also on employment. The long-term effects on employment are less clear than the effects on real growth. An example makes this clear. The Dutch petrochemical industry has been very successful in implementing new plants in which productivity is much higher than in the old installations, but only a few people are needed to operate the whole plant. The effect of these process innovations is a decrease of employment. If, however, this industry had been non-innovative-petrochemical firms in other countries might have implemented new installations that were more competitive than the 'old' Dutch ones. As a result the Dutch firms would lose market share and employment. The effect of innovation on employment can thus be positive or negative. It is therefore useful to make a distinction between the impact of innovation on employment in an industry in general (that is in all countries) and the impact of innovations specific for the industry in one country. In the general case the effect depends on the 'age' of the industry. In a new industry many product innovations occur and especially when a dominant product design is established (we come back to this later) sales and employment grow strongly. In more mature industries the emphasis is on process innovations that tend to reduce employment. To the extent that productivity still increases in the 'old' industries, economic resources can be allocated to 'new' industries. The balance between old and new industries and the nature of innovation (product or process) ultimately determines the overall employment effect. Given this general pattern it is clear that if one country performs better in terms of innovation than the others a positive additional effect on employment is to be expected. In this chapter we will pay more attention to the impact of innovation on exports and productivity than on employment. The reasons are that the available space in the book is limited and the impact of innovation on export and productivity is more clear.

Before we give insight into the economic significance of the effects of innovation one needs to understand the problems of measuring innovation activities themselves. The beginning of this chapter is therefore devoted to problems of measurement of innovation and only thereafter do we turn to the measurement of the economic impact of innovations.

This chapter begins with a short overview of how economic science has been treating technology in the past. The answer is that the process of innovation has been neglected and that one concentrated on technical change as being the economic effect of innovation. A subsequent section is therefore pointing out the difference between innovation and technical change. The following section focuses on the problem of how to measure innovation. It then examines technology's effects at the national, industry, and firm level. Thereafter the focus is on empirical evidence of technology's importance for the economy as a whole, and for industries and firms. The chapter concludes with a presentation of the main empirical facts on the creation and use of new technology.

2 A BRIEF HISTORY OF TECHNOLOGY IN ECONOMIC SCIENCE

For the classical economists, such as Adam Smith, technology was an integral part of the economic system. Well known is his pin factory, where the introduction of machines (changes in process technology) made possible a division of labour which resulted in higher labour productivity. From the end of the last century onwards, economic science focused its microeconomic analysis of production on an analytical understanding of the firm's choice process for a given set of technological alternatives. The influence of prices on this choice process, mediated through the market mechanism, dominated the analysis. An example may illustrate the way of modelling technology, until the second half of this century.

Let us assume that a market exists where a homogenous product P is fabricated with the use of process technology T_1. No economies of scale in production exist and the unit production cost C(P) only depends on how efficiently a firm uses technology T_1. Over a long period of time new process technologies may become available, but currently one alternative production technology T_2 is on the market. There are two production factors, labour and capital. T_1 is a relatively labour-intensive technology and T_2 a relatively capital-intensive technology. For example, the production of 100 units of product P with the use of T_1 requires two units of labour and one unit of capital, while with T_2 this is one unit of labour and two units of capital. If capital and labour costs are equal for all firms in the market,[4] all firms will choose the technology which minimizes unit cost C(P), and competition merely depends on how efficiently each firm uses the optimal technology. The creation of technology is an exogenous factor and those firms will survive in the long run which are best informed about short-term price changes in the factors of production and about the cost structure of the use of process technologies which are, or will soon be, available.

Ayers (1978) gives a critical but very clear review of neoclassical microeconomics' preoccupation with price theory, and stresses at the same time the (more) important role of technological change. In terms of the foregoing example Ayers' criticism would be focused on the questions how firms create process technologies and why one firms is actually adopting a specific technology and another is not.

At the macro level the relation between investment and economic growth was the core issue, but no attention was paid to R&D as a special form of investment. The creation and implementation of new technology was not an object of inquiry. Economic theory saw new technology as an exogenous variable until the early 1950s.

There were a few exceptions. Best known is the work of Schumpeter. In Chapter 1 we presented Schumpeter's four classes of innovation. Scherer (1984, vi) sees three substantive premises in the writings of Schumpeter between 1911 and 1942:

1. The notion that technological innovation gives capitalistic economies their peculiar dynamics through a process of 'creative destruction'; old products and industry structures are repeatedly displaced or altered by new forms.
2. The growth of real income per capita was not for the most part attributable to increases in population, money supplies or land resources, or to governmental actions, such as tariff reform, but to technological progress. This technological progress was established by private firms by their organizational and strategic abilities.
3. A market situation of strong competition was not conducive to the development of such organizational and strategic abilities to create innovations. The expectation of a monopoly position or even the possession of one was required.

Another, but less known, exception is formed by the 'old institutionalists', especially Veblen, who is regarded as one of the founders of this old institutionalism and saw institutional and technological change as the cornerstones of the whole economic process (see, for example, Veblen 1958). The earlier mentioned Ayers can be seen as a member of this school of economic thought.

The turning point, according to many authors, from 'the economics of innovation', were the findings of Solow and others in the early to mid 1950s about the apparent influence of technology on the long-run growth of GDP in the United States of America.[5] According to these authors only one-third of the observed growth in output could be explained by the growth in the traditional inputs labor and capital. The unexplained, the 'residual' in the regressions, accounted for two-thirds of the growth of GDP. This residual was afterwards interpreted as reflecting factors which improved the efficiency of use of capital and labor, and technological change was seen as the most important underlying force. In the wake of these studies, which had quite an impact on the community of economic scientists, various authors tried to test the model at the level of industries or tried to extend the original model by incorporating factors which could explain the increase of efficient use of production factors.

The emphasis, however, was on the impact of technology on economic growth through the use of new technology, not on the processes of creation and implementation of innovations. Here we will call the economic impact of the use of new technology 'technical change', while creation and implementation of new technology is called innovation.

3 THE DIFFERENCE BETWEEN INNOVATION AND TECHNICAL CHANGE

The concepts innovation and technical change have a different meaning, although not all studies in the economics and management of innovation make this distinction. I prefer to make a clear distinction between the two concepts and to use it consistently throughout the book, because a proper assessment of the technological performance of firms, industries, and whole economies cannot be made without a clear distinction between innovation and technical change.

Innovation has been presented as the effort of individual firms, although we will see later that firms may cooperate to create innovations. Each innovation brings about a technical change and, of course, if many innovations by firms occur in an economy, they cause a significant technical change. But not every technical change is an innovation. For example, when a firm purchases a new machine this implies a technical change for the purchasing firm.[6] In fact each change in a firm's technology, be it within its set of products, its production processes, or its component technologies, is a technical change for the economy. Clearly innovation somewhere and at some time in the economy is required to cause the kind of technical change meant here. Imitation and diffusion determine how much technical change is caused by a specific innovation. Technical change is therefore the more encompassing concept, it includes innovation, imitation, and diffusion.

4 MEASUREMENT OF INNOVATION

In the previous chapter the emphasis has been on the nature of innovations and the nature of the innovation process. In fact we discussed innovation as a process and as an output of that process, and we looked at different kinds of output and different kinds of activity within the process. Here the focus is on the measurement of innovation as an input and an output. The production function approach (see the introductory paragraph of this chapter) focuses on productivity as the ultimate and crucial effect of R&D. Some other approaches include

Table 2.2 Structure of the R&D costs in the business enterprise sector in OECD countries (percentages)

	Personnel expenditure			Other current costs			Total current costs			Capital expenditures		
	1981	1985	1989	1981	1985	1989	1981	1985	1989	1981	1985	1989
United States	46.0	42.9	39.9	54.0	57.1	60.1	100.0	100.0	100.0	..	..	..
Canada 1987 rather than 1989	46.7	48.0	48.8	40.2	36.2	37.3	86.8	84.1	86.0	13.2	15.9	14.0
Japan	44.2	40.7	39.4	39.8	42.9	44.9	84.0	83.5	84.4	16.0	16.5	15.6
Belgium	..	..	..	..	..	..	..	..	..	..	..	..
Denmark	64.6	58.2	60.8	25.5	22.7	25.5	90.1	80.9	86.3	9.9	19.1	13.7
France	55.6	53.7	48.7	37.0	38.0	41.2	92.5	91.7	89.9	7.5	8.3	10.1
Germany	57.8	57.4	57.8	30.8	30.7	30.9	88.6	88.2	88.8	9.7	10.9	11.1
Greece 1986 rather than 1985, 1988 rather than 1989	..	50.1	49.1	..	19.1	18.1	76.3	69.3	67.2	23.7	30.7	32.8
Ireland	48.3	44.1	47.7	28.0	31.9	30.5	90.4	76.0	78.2	9.6	24.0	21.8
Italy	48.9	42.1	40.8	41.5	46.4	47.6	90.5	88.4	88.4	9.5	11.6	11.6
Netherlands	56.9	50.6	49.3	33.6	37.1	38.2	83.8	87.7	87.5	16.2	12.3	12.5
Portugal 1982, 1986 and 1988	65.0	57.0	51.0	18.7	21.6	23.4	88.6	78.6	74.4	11.4	21.4	25.6
Spain 1988 rather than 1989	69.1	53.2	51.4	19.5	29.1	26.7	91.5	82.3	78.1	8.5	17.7	21.9
United Kingdom	44.9	42.2	42.2	46.6	47.7	46.9	87.8	89.9	89.0	12.2	10.1	11.0
Austria	59.9	56.9	55.6	27.9	29.0	32.5	87.8	85.9	88.0	12.2	14.1	12.0
Finland	55.5	51.6	48.0	32.6	37.4	38.3	88.2	89.0	88.3	11.8	11.0	13.7
Iceland	61.0	54.6	52.8	31.2	32.6	31.8	92.2	87.1	84.6	7.8	12.9	15.4
Norway	55.9	51.6	51.7	34.8	37.3	37.6	90.7	89.0	89.0	9.3	11.0	10.6
Sweden	55.1	47.0	49.1	34.5	40.5	40.9	89.6	87.5	90.0	10.4	12.5	10.0
Switzerland	..	..	..	..	..	..	90.9	..	..	9.1	..	..
Turkey												
Australia 1986 rather than 1985	60.8	50.8	48.9	26.1	35.9	36.2	86.9	86.7	85.1	13.1	13.3	14.9
New Zealand	..	..	..	..	..	..	..	..	..	..	..	..
OECD Median	**55.7**	**50.8**	**49.1**	**33.1**	**36.2**	**37.3**	**89.6**	**87.1**	**86.3**	**10.1**	**13.1**	**13.7**

Source: OECD (1994, 163)

more direct relationships between R&D, innovations, and commercial results obtained with those innovations. The productivity of R&D remains one of the main topics, however.

The measurement problems concerning innovation can thus be divided into three categories: inputs of the innovation process, direct outputs of the innovation process, and the economic effects of innovation output on the growth of the firm and the growth of the economy. We begin with measurement problems concerning inputs and direct outputs of the innovation process at the firm level.

R&D, and its various subactivities are the inputs to the innovation process. They are usually measured in money terms, and called 'R&D expenditure', or measured in units of labor input, usually manyears of R&D labor. Labor is the most prominent input in most innovation projects, but sometimes expenditure on research equipment, buildings, test grounds, and material can be substantial. According to Soete (1979, 320): 'R&D employment is considered here to cover only the number of scientists and engineers. Other commonly considered R&D input cost are 'other supporting personnel', 'material and supplies', 'durable equipment', 'land and buildings' and 'other R&D costs', such as energy, water, maintenance, etc.'

Table 2.2 gives figures on the relative cost of the main inputs to the R&D process in general, without looking at the earlier discussed distinctions between kinds of innovation output and kinds of innovation activity.

In an economic sense the largest part of the expenditure on R&D is an investment. The basic characteristic of an investment is that it must be paid now, but will yield a return later. As we discussed earlier most of the investment is done during the development activity of an innovation project. One only knows the exact investment in R&D once a project has been completed; it is the accumulated R&D expenditure over the total period from the beginning of the project until completion.[7] In quite a number of cases the time of completion may in reality mean the moment of cancellation of the project. Only for successful projects, that is the introduction of a new product on the market, the in-house application of a new process, or the use of newly developed 'generic' technologies in product or process innovation projects, can one really speak of completion time. In our further discussion of R&D productivity we begin with a single project and only thereafter do we address the question of R&D productivity of a whole firm, or even larger entities.

The input of an R&D project is defined here as the accumulated R&D expenditure of that project. What is the output? From an economic point of view the only output that really matters is the innovation; failures and inventions which are not being commercialized are not outputs.[8] A possible indicator for the output of an innovation project is therefore the number of innovations. Here a problem arises. Is a project with one innovation as a result less than a project with two or more innovations? Is a project resulting in a new gear system for a mountain bike less or more important than a project resulting in a new DRAM chip? An ordinal scale (that is, number of innovations) seems to be acceptable only in the situation where the kinds of innovation are comparable. If one project comes up with three innovations concerning improvements of an existing product with which the market share can at best be kept, and another project with a radically new product with which a whole new market can be opened up, we clearly talk about different kinds of innovation.

It is therefore important to compare similar kinds of innovation. But what is similar? On the one hand one needs a criterion for newness, on the other hand a criterion for economic impact. Both kinds of criterion are difficult to assess in a quantitative way for people from the innovating firm, let alone for some outside analyst. In many cases the degree of newness will be positively correlated with the economic benefits of an innovation project. But this is not true for every project; for example, a number of minor product improvements may have substantial market effects, while a radical innovation can fail in the market. Let us for the time being assume that the degree of newness correlates perfectly with the size of the benefits yielded by the project.

Ideally one ought to possess reliable estimations of the total costs and benefits of a project. In some projects, like basic research projects, the benefits cannot be calculated in terms of profits related to a particular market, because the goal of the project is not related to an immediate commercial application. Only for projects realizing a commercial application of a specific process or product (improvement) can estimations of profits in financial terms be made, at least theoretically. Similar to accumulated R&D expenditure at the input side we need accumulated innovation profits at the output side. We then have R&D costs and innovation profits, both in monetary terms. R&D productivity of a project can be calculated as profits divided by costs. The advantage of being able to calculate productivity in such a way is that an 'objective' ground has been established to compare different projects. An R&D budget of a firm could then be divided among potential projects on the base of expected net profits. It is worthwhile to examine the benefits and costs of a firm's innovation projects more carefully.

Let us define three types of project: projects without any direct commercial application which we will call basic research projects (P^b), development projects aiming at commercial application of a new or improved product or process, which are successfully completed (P^{cs}) or which are a failure (P^{cf}). Only for projects which lead to a commercial application of new technology should one calculate costs and benefits. For a successful project P^{cs}_i one can write:

$$B^n(P^{cs}_i) = B(P^{cs}_i) - C(P^{cs}_i) - f[C(P^{cf}_j) + C(P^b)] \qquad (2.1)$$

In which: B^n are the net benefits of the project.

B is the net present value of the accumulated commercial benefits of the innovation(s) stemming from the project.

C is the net present value of the accumulated innovation costs of that project, restricted here to R&D expenditure.

$C(P^{cf}_j)$ is the net present value of the costs of all the projects j which failed during the period in which project i has been undertaken.

$C(P^b)$ is the net present value of the costs of all the basic research project in that period.

f is the fraction of the costs of all unsuccessful projects and basic research projects assigned to project i.

As was said before, exact calculations of R&D expenditure can only be made after project completion. In practice cost estimations made at the beginning are adjusted during project execution when more knowledge is gained about the innovation requirements. The same holds for estimations of the benefits of an innovation.[9] Given such adjustments of estimations of costs and benefits of an innovation project, Figure 2.2 represents a common function of costs and benefits of an innovation project.

One difference between cost and benefit estimations lies in the time at which reliable estimations can be made. For R&D expenditure this is close to project completion. For benefits it is close to the time when commercial success of the innovation stops, for example when a new product for which estimations have been made is replaced by another new product in the market. The period over which benefits of the innovating firm must be calculated is therefore dependent on the time subsequent new products are brought on the market by this firm or by the other firms in the market. Competition in innovation complicates estimation of the profits of an innovation project, because it will depend on whether competitors are more successful with their innovations than you are. We come back to the competition aspects in Chapter 5.

An additional complication to the above analysis arises if one introduces different kinds of innovation. Different kinds of innovation make comparison of projects even more difficult. This might give rise to the use of other output indicators, as we will argue below.

From a management point of view it might even be more interesting to compare the R&D productivity of various teams inside the firm (and with teams outside) than the productivity of projects. A problem is that teams may change in composition. What could be done is a comparison of the productivity of the various R&D departments. Obviously only large companies have a number of departments (for example at each division).

We now turn again to the effects of different kinds of innovation on measurement. Before we do, it must be understood that different kinds of innova-

Figure 2.2 Project cumulative cash flow diagram

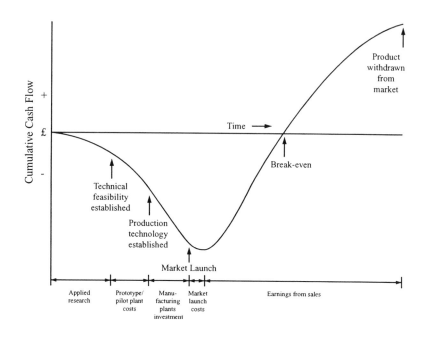

Source: Twiss (1992, 147)

tion can be compared if reliable estimations of the net profits of each project can be made. The relative difficulty of estimating benefits of innovation projects depends on the kind of innovation, however. For some innovations, like routine improvements, the benefits are much easier to estimate than for other innovations, such as a radical product innovation aiming at an entirely new market. Consequently introducing different kinds of innovation in the analysis implies an aggravation of the measurement problems.

Comparison at the firm level

In the previous section it has been shown how difficult it is to make estimations of costs and benefits of an innovation project. We stated there that comparison of projects is best made when benefits are calculated in money terms. Such a comparison at the firm level is complicated by the fact that many projects carry costs but do not yield benefits. The successful projects must be charged with the R&D costs of the unsuccessful projects to get correct estimates of the costs of the former (as was shown in equation 2.1). On the other hand, the results of a successful project at one point in time may be positively influenced by the results of concurrent projects or previous projects. Even uncompleted projects from the past may have a positive effect on current projects (learning from one's mistakes). We come back to the interrelatedness of projects in later chapters.

Despite all these problems one can argue that part of the errors in measuring project productivity can be avoided by looking at the R&D productivity of the firm, because mistakes in estimating costs or benefits of single projects may be averaged out. What is the overall R&D productivity of a firm? The productivity of a firm in year t can be defined as the accumulated benefits derived from a number of innovation projects which have been completed in year t divided by the total R&D costs of these projects. The 'general' R&D productivity of the firm can be defined as the average R&D productivity over a number of years.

Calculating the accumulated net benefits of a given set of innovation projects is very difficult because, as we argued above, estimating costs and benefits for a single project is problematic and interdependencies exist between projects. Nevertheless, managers use such estimations, at least in a rough way, in order to assess the economic value of a redistribution of a given R&D budget among alternative projects, or of a change in the overall R&D budget of the firm. In practice firms quite often work with fixed budgets and we will see in Chapter 5 that firms do not always react in a predictable way in terms of changes in the R&D budget when the pressure of competition changes. We will see in Chapter 3 that the allocation of resources over a number of projects is made more complex because the strategic importance of different types of project may vary, regardless of the net benefit. One such strategic element is the impact one innovation project (which may have a net direct loss) has on the cost or benefits of several subsequent projects.

Because of the measurement problems there have been relatively few economic studies measuring R&D productivity of the firm. In practice additional difficulties may arise because of a lack of data. Quite often there are only data available on sales per division, not per product group, or when figures are available per product group, no distinction is made between old and new products. For a radical product innovation, opening up a new market (and thus not replacing old products) measurement of benefits ex post[10] is usually easier than for incremental product innovations, which are made continuously and where a number of improved products overlap a number of old generation products. Things are just the other way around for (ex ante) estimations of benefits of incremental and radical innovations.

Comparisons at the industry level

At the industry level the innovation performance of the firms can be compared too.[11] Here the previously mentioned data problem arises again. We do have statistics on R&D expenditure of the firms, but estimations of the commercial benefits are normally not available. Consequently we cannot calculate R&D productivity. Instead R&D intensity is used, which merely indicates how many inputs on average are used by the firms in the industry relative to their size. Occasionally survey data are available with figures on number of innovations, but most of the time we have to rely on patent statistics of national and international patent offices (see chapter 6 for further information on patents). In such cases R&D expenditure at the industry level can be compared to the number of patents in a specific year. It must be realized, however, that a time lag exists between the R&D expenditure resulting in patentable new

Table 2.3 Commercial benefits of product innovations in some European countries in selected sectors

NACE Sectors	Countries: Name of sector	NL.	N.	DM.	A.	G.
		Share of innovative products in 1992 sales[a]				
10–14, 40–41	Mining etc., Food industry	22.1	24.9	NA	22.7	36.0
15–16	Electricity production	32.0	44.7	48.0	20.1	34.0
17–18	Water supply	38.6	33.1	47.0	48.8	43.0
20–22	Metals production	27.4	21.7	24.0	29.7	30.0
23–25	Chemical industry	30.6	27.4	27.0	31.6	51.0
26	Man-made fibres industry	28.1	23.7	23.0	28.0	1.0
30,32	Mechanical engineering	47.0	55.5	37.0	45.5	77.0
31	Metal products	43.2	51.5	29.0	40.7	46.0
33	Office machinery	42.0	55.8	38.0	48.6	51.0
34,35.3	Electrical engineering	45.6	31.0	38.0	43.2	60.0
35(excl.35.3)	Motor vehicles and parts	35.7	46.1	40.0	10.0	36.0
36	Other means of transport	38.5	45.8	41.0	50.3	66.0

[a] Innovative products are defined as products new for the firm.

Source: derived from Kleinknecht (1996, table 3)

knowledge and the actual reception of the patent. In addition, and more importantly, we will see later on that patents are an imperfect indicator for the commercial output of innovation activities.

Measuring the economic impact of innovation

Traditionally economists have been occupied with the measurement of technical change and not with the measurement of innovation as such. The main way to measure technical change has been by use of productivity growth figures. In fact the earlier mentioned studies of Solow and others about the output growth of the US economy in relation to the growth of labor and capital inputs followed this approach. In the theoretical section we come back to this issue in terms of 'production functions'.

It is important to note that instead of measuring R&D productivity, we now talk about linking R&D (or innovation) to labour productivity or 'total factor productivity' at various levels of analysis. The reason for this is partly that no adequate data are available about the direct commercial effects of single innovation projects or of the total R&D expenditure of the firm. Consequently we have few reliable estimations of R&D productivity.

Recently a few studies have been carried out in which survey questions have been asked about the percentage of sales obtained by new products. In this way the R&D per year can be linked to the commercial results of the new products stemming from the R&D activity.[12] Table 2.3 provides a few results.

5 EMPIRICAL EVIDENCE ON THE ECONOMIC EFFECTS OF INNOVATION ACTIVITIES

Now we know about innovation measurement we have a sufficient base to discuss how the importance of innovation for the economy is assessed.

Most economists agree on the fact that without innovation a nation's economy will lose its competitive position against nations that do create many innovations. But we must be careful to distinguish what determines the competitive position of a country? First of all the competitive position of its private firms in the domestic markets and in the international markets is decisive. A country can be said to be 'competitive' if 'a sufficient number' of its firms at least keep their market shares *vis à vis* foreign firms. There are no equivocal quantitative criteria on what is a sufficient number, but it is obvious that if all of a country's firms lose market share (and profits at the same time) the economy will sooner or later decline. The matter is, however, more complex. For example it is possible that the same country is very attractive to foreign firms. If many firms invest in the country the economic situation might change. Nevertheless one agrees that consistently poor performance of domestic firms will weaken a country's competitive position. Under this assumption the main question is whether innovation is necessary to change the situation.

Generally it is not questioned that a relative high performance of innovation of a country's firms will improve its position. But it is also possible that a country's firms have modest innovation performance and keep their competitive position by purchasing new technology on the market. A theoretical issue we go into later is to what extent an effective purchasing of new technology requires a 'sufficient amount' of innovative domestic firms. Basic to this question are the relations between innovating firms and their customers and suppliers. Another element is the quantity and quality of a country's knowledge infrastructure. We deal with this matter in the chapter on technology policy, but here it is submitted that a country's firms may perform more efficiently or effectively when that infrastructure supports innovation and the efficient use of technology.

When firms are competitive through innovation, they may improve the productivity of the country of residence in two ways. First, process innovations increase the productivity of these innovating firms. Second, product innovations used by other firms increase the productivity of these using firms.[13] This is why, despite the more complicated picture shown above, productivity increase in the economy is seen as the main economic effect of innovation.

Figure 2.3 The contribution of TFP (R&D) to GDP

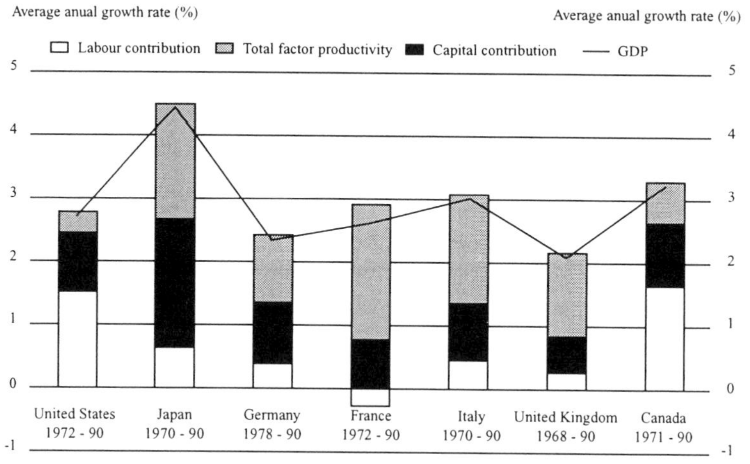

Source: OECD (1996, 26)

Figure 2.3 provides some figures for OECD countries about productivity growth and (indirectly) the effect of technical change. In a so-called 'aggregate production function' the output (growth) of a country is linked to the growth of the stock of the traditional production factors labour and capital and to a remaining factor called 'total factor productivity' (TFP).

At the level of the OECD countries a positive relationship has been estimated between R&D per employee and the total factor productivity level in the business sector in the period 1970–90. The growth in both figures was also positive correlated (OECD, 1996).

Effects at the meso (industry) level

The models relating productivity and R&D have also been applied at the industry or meso level. Other approaches focus on the relation between R&D and export. The approach with the greatest empirical tradition is 'Industrial Organization' where R&D is related to other industry characteristics, such as average firm size and concentration. In this tradition one is also concerned with the sources of innovation and thus with the factors that determine the volume and nature of R&D expenditure. Ultimately one wants to explain the innovativeness of the industry, which could be measured as the average R&D productivity of the firms.

At the industry level a significant positive correlation between R&D and productivity remains, but there are considerable differences among sectors. Most of the empirical studies have been carried out with data about the US economy.[14] Hall and Mairesse (1996) have undertaken statistical tests for France. In a study on Japan, Odagiri and Iwata (1986) found that over the period 1966–82 R&D expenditure as a ratio to value added positively influenced total factor productivity of the 135 to 168 Japanese manufacturing companies studied. Interindustry differences do occur, however.

Two sorts of question can be asked with regard to the production function studies. The first one is theoretical and concerned with the relevance/appropriateness of the assumptions underlying the model. According to Griliches (1994), who did much empirical work in this area, one could indeed question the assumptions of the model and he suggests an extension of the model. We come back to this topic in the theoretical section of this chapter. The second type of question focuses on the availability and reliability of data. The main conclusion of the Griliches article (Griliches, 1994) is that the correlations between R&D and productivity in rapidly changing industries, such as the computer industry, do not reflect what is going on in practice. A main reason is that no industry-specific deflators are available which means that, in the case of the computer industry, the drastic price decreases in computer products and their components are not taken into consideration in the statistical tests. As a result real output and productivity in the industry are strongly underestimated. In addition it can be said that in the last three to four decades the service sectors have gradually replaced manufacturing as the largest sector type in the economy, and even within manufacturing the importance of services has increased considerably. It is well known that estimations of productivity in services are very difficult to make.

As has been argued before, R&D leads to innovation, which in its turn leads to improvement of a firm's competitive position. In a world with increasing international competition in markets, a firm's competitive position is best reflected in its performance in international markets. Export performance of domestic firms and industries is used as a measure of this performance on international markets. New products and production processes give a firm the opportunity to compete on international markets. The creation and use of new technology in an industry is therefore expected to be positively related to its productivity (in the case of new processes) and export performance (product or process). The success of the Japanese economy is well known in these respects. It is therefore worthwhile to mention an empirical study of the above relations for this country.

Ito and Pucik (1993) studied three factors influencing the export performance of Japanese manufacturing firms: R&D spending, domestic competitive position, and firm size. They conclude that a firm's export ratio is related to the size of the firm, but not to the firm's and the industry's R&D intensities. We come back to this point when discussing a Dutch study on the topic. A further conclusion is that follower firms are characterized by higher export ratios than market leaders. This suggests that a relationship exists between the pattern of domestic competition and the international competitiveness of Japanese firms. Technology may be important to explain exports, but the kind of technology strategy matters.

In the economics literature a series of empirical studies have been undertaken the last one or two decades in which trade performance is related to technology. In a review of the literature Soete (1987, 103) states: 'Technology has undoubtedly emerged as one of the most important factors in explaining international trade flows.' His own empirical tests relate a country's share among OECD countries in total export of a product, expressed in the country's total share of OECD exports (the so-called revealed comparative advantage index), to a number of explanatory variables, including the country's share of total patents granted in the US for each product.[15] In comparison to most studies in this field Soete uses an innovation output indicator instead of the input indicator R&D (intensity). The main conclusion of Soete's research is that technological performance related to the international market (approximated by patents granted for the US market) gives a forceful explanation of relative export performance within the OECD. More specifically the highest influence of the technology factor was found in industries with a high patent intensity, and non-significant results or a negative relationship were found in industries which are bound to be strongly dependent on the availability of natural resources or homogeneous labour and capital. Or, as Soete (1987, 124) formulates:

> One might indeed expect that any increase in a country's relative (i.e. as compared to its competitors) technological performance will be more rewarding in terms of relative export performance, in technology-intensive industries than in non-technology-intensive industries.

Brouwer and Kleinknecht (1993a)[16] have undertaken a study, similar to the study of Soete, for the Netherlands. At the meso level 41 Dutch industries have been compared with similar sectors of the main trading partner of the Netherlands, that is, Germany. For the years 1983 and 1988 the relation between R&D intensity and export performance has been estimated,[17] controlling for the effects of some other variables. The main result is that a positive correlation can be found between the development of the R&D intensity of 41 Dutch industries in 1983–88 and their exports. This significant influence disappears, however, when exports of Dutch industries are corrected for import penetration. Instead 'investment per employee' becomes significant. This can be interpreted as the adoption of capital-embodied technological change. Apparently, Dutch industries improve their export performance by increasing R&D intensity, while their position against import penetration is defended by purchasing technologically advanced machines and equipment.[18] Worth mentioning is that growth of domestic sales leads to less exports (and more import penetration), but also to an increase of R&D intensity in the sector. This translates itself into an improved export intensity later.

Effects at the firm level

Positive correlations between R&D expenditure and productivity growth have also been found at the firm level, although relatively few studies have been carried out at this level. Figure 2.4 presents data on R&D intensity and the annual growth rate of total factor productivity of 15 pharmaceutical firms, as provided by Bean (1995, 26).

The relationship between export and R&D has also been investigated at the firm level. The study of Brouwer and Kleinknecht includes data at the firm level. At this micro level it has been analyzed which factors explain export's share of total sales of 2.165 Dutch firms in 1988. The effects of R&D have been controlled by a number of variables of which scale effects, sector-specific features, regional location, R&D cooperation with foreign partners and the purchasing of advanced machinery and equipment had a systematic influ-

*Figure 2.4 The correlation between R&D and TFP growth in the
pharmaceutical industry*

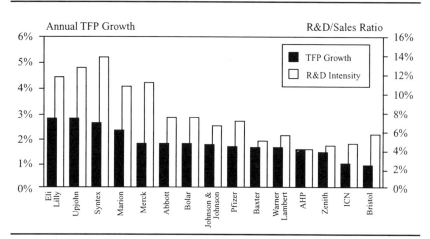

Source: Bean (1995, 26)

ence.[19] R&D itself kept its significance, though only for product-related R&D activity. Process-related R&D activity did not have a significant positive influence on export intensity.

Studies at the firm level open up a whole new line of thought. The question becomes 'why are some firms in sector X of a country more competitive than other domestic firms in X?' In Chapter 4 we discuss success and failure factors of individual innovation projects. In Chapter 3 some management and organizational issues at the firm level are discussed.

A disadvantage of the studies relating R&D and export performance at the industry level is that they do not use a theory explaining how an increase of R&D may lead to an increase in exports. In contrast the Industrial Organization approach does use models which link R&D to innovation performance at the firm level in combination with factors operating at the industry level, such as competition and technological opportunities. Some of these models will be discussed in Chapters 4 and 5. Here we give a short picture of this approach.

Industrial Organization takes the view that innovation performance of firms in an industry (on average) can be explained by the nature of the competition process in that industry and other variables characterizing the industry as a whole. Quite often the complex and dynamic nature of the competition process is reduced to a static reflection, the so-called market structure. This market structure can be defined as the number and size of the firms in the industry in a particular time period and the entry possibilities for new firms. The question is 'which market structure gives the best innovation results for

Table 2.4 Changing market share ranking in electronic components

	Valves	Transistors	Semiconductors							
	1955	1955	1960	1965	1975	1980	1983	1984	1985	1988
RCA	1	7	5	6	8	-	-	-	-	-
Sylvania	2	4	10	-	-	-	-	-	-	-
GE	3	6	4	5	-	-	-	-	-	-
Raytheon	4	-	-	10	-	-	-	-	-	-
Westinghouse	5	8	-	-	-	-	-	-	-	-
Amperex	6	-	-	-	-	-	-	-	-	-
National Video	7	-	-	-	-	-	-	-	-	-
Ranland	8	-	-	-	-	-	-	-	-	-
Eimac	9	-	-	-	-	-	-	-	-	-
Lansdale Tubes	10	-	-	-	-	-	-	-	-	-
Hughes	-	1	9	-	-	-	-	-	-	-
Transitron	-	2	2	9	-	-	-	-	-	-
Philco	-	3	3	8	-	-	-	-	-	-
Texas Inst.	-	5	1	1	1	1	1	1	2	5
Motorola	-	9	6	2	5	3	2	3	4	4
Clevite	-	10	7	-	-	-	-	-	-	-
Fairchild	-	-	8	3	2	7	-	-	-	-
GI	-	-	-	4	7	-	-	-	-	-
Spraque	-	-	-	7	-	-	-	-	-	-
Nat. semi-cond.	-	-	-	-	3	2	7	6	8	-
Intel	-	-	-	-	4	5	8	7	7	10
Rockwell	-	-	-	-	6	-	-	-	-	-
Signetics	-	-	-	-	9	4	6	9	10	6
AMI	-	-	-	-	10	-	-	-	-	-
NEC	-	-	-	-	-	6	3	2	1	1
Hitachi	-	-	-	-	-	8	4	4	3	2
Fugitsu	-	-	-	-	-	9	9	8	6	7
Toshiba	-	-	-	-	-	10	5	5	5	3
Mostek	-	-	-	-	-	10	-	-	-	-
Matsushita	-	-	-	-	-	-	10	10	9	8
Mitsubishi	-	-	-	-	-	-	-	-	-	9

Source: Twiss (1992, 36)

the firms on average?'. It is also recognized that the innovation performance of one or a few firms might change the existing market structure. If one firm grows very fast due to a successful new product the size distribution of firms changes and in the extreme all other firms will be driven out of the market, hence a monopoly remains. Empirically, changes in the market structure can be identified by looking at the market share of firms.

In markets with a rapid technological change it is not uncommon that market shares of existing firms change drastically over the years, and in addition firms exit the industry and new firms enter. An example is given in Table 2.4.

Figure 2.5 Hypothetical production function

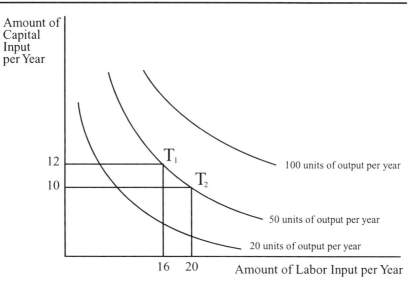

T$_1$= process technology 1
T$_2$= process technology 2

Source: Mansfield (1977, 13)

6 A THEORETICAL EXPLANATION

As we said before, economics used to focus on measuring the economic effects of innovation, that is, technical change, and not on the number of innovations or other output measures of the R&D process itself, such as patents. We begin with a theoretical analysis of technical change and then proceed with an explanation of the more direct relation between R&D and innovation outputs.

The production function

A production function relates output to a number of inputs, usually called factors of production. Inputs and output can be related in various ways. An example is the Cobb-Douglas production function in equation (2.2) in which the production factors capital and labour are perfect substitutes and these inputs have a constant elasticity to scale.

$$Q = A.C^a L^{1-a} \tag{2.2}$$

Figure 2.6 Technically efficient combination of labor and capital inputs per unit of output at three levels of technology

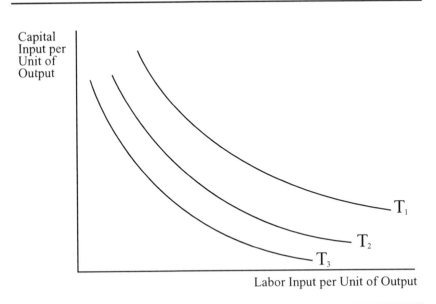

Labor Input per Unit of Output

Source: Mansfield (1977, 14)

Production functions have been used at various levels of analysis: production unit, firm, industry, national economy. At the level of the production unit the production function can be given a direct technical interpretation: the output of a specific product P can be realized with different combinations of the inputs L and C. We use Figure 2.5 for a further explanation.

The three combinations of inputs in Figure 2.6 represent three production techniques T_1, T_2, and T_3 which can produce the given amount of output 2Q per year of a specific product P. For other levels of output, such as 1Q or 3Q other amounts (and possibly other combinations) of labour and capital are required. Production functions indicate maximum production with a given combination of inputs.[20] The idea behind them is that a firm in the market for P can make a choice of the most efficient one from a set of available techniques, depending on the relative prices of labor and capital. When a firm uses T_2 and its market share remains the same (2Q) an increase in the cost of labor relative to the cost of capital equipment would make T_1 a more efficient production technique.

When returns to scale are constant, as in the case of the Cobb–Douglas production function, T_1, T_2, and T_3 would represent technically efficient production techniques for all levels of output. The production function can then be represented as efficient combinations of capital input per unit of output and labor input per unit of output (Mansfield 1977, 14).

Technical change can be defined as the shift of the production function over a given period of time. The more the function shifts downward the greater becomes the technical change in the economy. An example would be the shift of the production function for all the firms in the market for product P. The technical change thus analyzed can be so-called labor-saving, capital-saving, or neutral. When, for example, an industry is confronted with a rise of wages over a longer period of time, the firms may want a new production technique T_4 which uses less labor per unit of output than the currently available techniques. Economic development, in the form of relative prices of production factors, may thus give direction to technological innovations.

A disadvantage of the studies at the macro level is that the innovation results of all firms in an industry are reflected in a single production function for the whole industry or, when R&D is included as a separate production factor, that no relation is made explicit about production (function) and R&D. Note that we already have seen that in some industries innovations are used that have been created by firms in other industries. Hence the relationship between R&D and productivity may cover several industries.

A few final remarks are needed to place this kind of economic analysis into the framework gradually developed in the book. First, the production function is about process innovations, not product innovations. According to (among others) Pavitt (1984) and Nelson (1982) a majority of all innovations are product innovations, not process innovations. Other studies show that between 50 and 80 percent of all R&D activity is focused on product innovations or a combination of product and process innovations.[21] Second, it is implicitly assumed that all available production techniques are known to all firms and when process innovations occur it is not analyzed how they are created and by whom. This led Rosenberg (1976, 64) to the conclusion that 'If, in response to a change in factor prices, a firm has to commit resources to establish new optimal input mixes, should not the activity leading to the new knowledge be described as technological change and not factor substitution?' Or, to put it differently: as the generation and diffusion of knowledge in the economy is costly, the assumption that all production techniques in operation or available in the near future are known to all the firms is highly questionable.

The lack of a theoretical foundation

It has already been stated that many empirical models on the economic impact of innovation lack a real theoretical foundation.[22] We discuss some of the theoretical problems here. We begin with the use of production functions to measure the effects of R&D at the macro and meso level.

Such studies usually incorporate comparison of countries or industries. The comparison of innovation performance in an industry in two countries is difficult for at least two reasons. First, innovation performance as such is difficult to measure. Second, each industry usually consists of firms with and firms without innovations, and the countries may differ in this respect. The industry in country A may perform better than in B simply because in A relatively more firms undertake R&D. This argument underscores the fact that one should relate innovative behaviour of firms to innovation performance at the industry or country level. The production function approach may be very useful to analyse whether an industry or country is performing relatively poorly, but to be able to do something about it, one has to understand the links to firm behavior. This point comes back in studies like Soete (1979), who relates R&D inputs and direct outputs.

The limitations of patents given[23] as an output indicator is a better proxy for innovation performance than R&D. R&D intensity of industry X in countries A, B, C, and so on might reflect as much 'imitative behavior' of the firms, with respect to R&D budgets of competitors, as efficiency of carrying out innovative activity.[24]

One important factor, according to Soete, not included in the empirical tests, is the availability of science and technology personnel as an innovation input indicator.[25] This relates to the role of human capital, which can be seen as an additional production factor.

An explanation of R&D productivity of the firm?

When R&D is included as a production factor in the production function of the firm correlations between R&D and (total factor) productivity of the firm can be estimated. In this approach, R&D is assumed to lead automatically to new technology and it thus reflects the influence of new technology on the productivity of the firm. Not included in the approach are: (1) factors that determine when and how much a firm invests in R&D; (2) factors that determine the innovation performance per unit of R&D input, and (3) factors that determine the extent to which the use of new technology leads to maximum performance (the maximum output that can be produced with a given combination of inputs).

All three types of factors are examined in the management of innovation field. The first two factors are also studied in the economics of innovation. Here the nature of technology and the nature of the market play an important role. Important factors discussed in the following chapters are: the nature and growth rate of demand, the nature of competition, the appropriability regime, and technological opportunities. They all influence the costs and benefits of innovations. In this respect a clear distinction must be made between the costs and benefits perceived by the firm and the costs and benefits for society. Decisions by firms to invest in R&D are based on their expectations of the costs and benefits of innovation projects. If an innovation project is perceived as highly uncertain a firm may decide not to invest, while the use of such an innovation can be very beneficial for society as a whole. The extent to which innovations are created by firms in one industry and used in other industries is the topic of the next section.

One of the main problems of innovating firms is to get reliable estimations of the benefits of innovation projects that they want to start. These benefits basically depend on the costs and time it takes to complete a project; on the 'fit' between the characteristics of the innovation and the preferences of the customers; and on the behaviour of competitors. Successful completion of an innovation project is therefore no guarantee for commercial success. As we will see later, innovation capabilities must be accompanied by so-called com-plementary assets (Teece, 1986), such as efficient production and distribution facilities, which enable firms to appropriate innovation profits.

7   THE CREATION AND USE OF NEW TECHNOLOGY: TECHNOLOGY FLOWS BETWEEN INDUSTRIES

In this chapter it has become clear that the economic significance of innova-tion depends as much on the extent of creative activity as on the degree to which innovations are used in the economy. In this section we give some fur-ther evidence on the extent and nature of technology flows between creators and users of innovations.

Figure 2.7 Technology flows between US industries in 1974

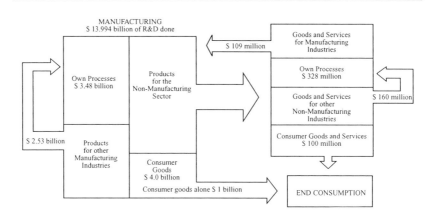

Source: Nelson (1982, 231)

From our discussion of R&D intensity in the introductory chapter it followed that industries differ considerably with respect to their innovativeness. Later in the book we will discuss various factors which influence an industry's R&D intensity, such as technological opportunities. Some industries have a low R&D intensity, because most firms are users of new technology, not creators. If such a 'low R&D intensity industry' in a country scores considerably lower that the same industry in other countries, we are talking about a competitive disadvantage. If, however, the R&D intensity is comparable with other countries, the industry is inherently producing few innovations and the criterion for competitiveness becomes the extent to which the firms in the industry use new technology created in other industries. In fact it is quite common that some industries create new technology, while other industries mainly use this technology created in other industries. We will give some data about new technology which flows from one industry to the other in the economy.

Technology flows between industries in the economy

At the industry level (sector level) it can be identified what the main sources of innovation are: the firms themselves, other firms, the scientific community, foreign countries. Here we limit the empirical discussion to the sectoral pattern of technology flows between creators and users of innovation. Nelson (1982) and Pavitt (1984) have undertaken studies for the US and the UK respectively in which the number of innovations produced and used by each industry were identified.

On the basis of specific information for the year 1974 of the R&D activity and patents of 443 large firms in the US, specified according to 4,274 separate lines of business, Scherer was able to aggregate innovations to 41 sectors form which the innovations originate and 53 sectors in which the innovations are used. Figure 2.7 shows that 95 percent of all industrial R&D is performed by manufacturing industry, while 50 percent of the R&D in manufacturing is devoted to new products for the non-manufacturing sector in the economy.

Most non-manufacturing sectors can therefore be said to be users of new technology. Within manaufacturing the picture is more diverse. Nelson distinguished different categories of sectors:

1. sectors with a balance between the amount of R&D spent and the amount of R&D 'received' (for exemple rubber and plastics);
2. sectors which receive much more R&D than they spend (for example, printing and publishing, lumber and wood, ferrous metals, textiles);
3. sectors which spend much more on R&D than they receive; this last category can be subdivided into:

Table 2.5 The pattern of production and use of innovations in some selected industries in the US in 1974

	R&D of sector	Users of R&D results: Own sector	Other sector	Cons.	R&D at suppls
Food and tobacco	444.9	278.2	23.6	143.1	245.0
Textile products	179.3	128.4	32.7	18.2	122.4
Apparel & leather	55.5	16.5	1.3	37.7	73.8
Organic chemicals	297.2	163.3	119.8	14.1	207.6
Pharmaceuticals	557.3	71.0	24.3	462.0	95.3
Rubber & plastics	419.8	203.0	104.9	111.9	470.0
Metal products	552.7	127.7	318.4	106.6	270.3
Engines & turbines	282.2	38.9	203.7	39.6	56.9
Electr. components	594.9	386.4	162.3	46.2	446.4
Motor vehicles	1518.0	158.8	13.3	1345.9	308.1

Source: own calculations based on table 2 of Nelson (1982, 232–41)

a. sectors where own R&D is embodied in products for a few using sectors (farm machinery industry, aircraft, missiles);
b. sectors where own R&D is mainly embodied in consumer products (food and tobacco, apparel, paper products, pharmaceuticals);
c. sectors where the R&D results in products for a broad array of consumer sectors (organic and other chemicals, rubber and plastics products, fabricated metal products, computers).[26]

Table 2.5 presents some examples of sectors from Nelson's study. For each sector the amount (in million US dollars) of R&D expenditures is given which is subsequently divided among the users of the innovations derived from that R&D: households (final consumers), other sectors and the sector itself (in case of process innovations).[27] In addition the expenditure on R&D in 'supplier' sectors is presented, which is the source of innovations used by the sector. The more a sector directs its R&D towards product innovations and not process innovations (own use), the less it will be correlated with productivity increase in the sector.

Pavitt (1984) also presents data on sectors which produce innovations and which use them. Although various important differences exist between this study and the one of Nelson, similar patterns of production and use occur, according to both authors. While Nelson examined supplier sectors as an input source to an innovating sector's production facilities (in the form of machines, parts or materials) Pavitt studied sources of information inputs to

Figure 2.8 The basic flows of information

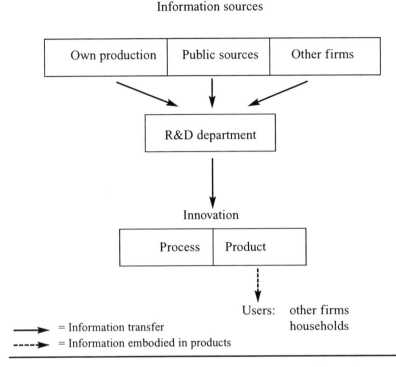

Information sources

| Own production | Public sources | Other firms |

R&D department

Innovation

| Process | Product |

Users: other firms
         households

⟶ = Information transfer

---⟶ = Information embodied in products

the innovation process of the producing sector, or one could say inputs to the R&D department of the firms in the sector. The information and technology flows for an innovating firm in a sector may thus be depicted as in Figure 2.9.

Technology relations between firms

The technology flows between different industries we presented suggest that firms in the user industries are purchasing the new technology from firms in the innovating industries. Also innovating firms purchase new technology from supply sectors. Empirical studies show that on various occasions the technology relationship between creator, customer and user is more than a simple contract relationship. We present some figures here and go into a deeper understanding of such relations in Chapter 3.

Customer–supplier relationships play an important role in practice. In many cases firms operating in a particular industry buy so-called intermediate products from suppliers and sell their own products also as intermediary products to their customers. At the beginning of the chapter it was shown that

Table 2.6 Summary of functional sources of innovation

Innovation developed by

Innovation type	User	Manufacturer	Supplier	Other	NA (n)	Total (n)
Scientific instruments	77%	23%	0%	0%	17	111
Semiconductor and printed circuit board process	76%	21%	0%	12%	6	49
Protrusion process	60%	10%	0%	0%	0	10
Tractor shovelrelated	6%	94%	0%	0%	0	16
Engineering plastics	10%	90%	0%	0%	0	5
Plastics additives	8%	92%	0%	0%	4	16
Industrial gas-using	42%	17%	33%	8%	0	12
Thermoplastics using	43%	14%	36%	7%	0	14
Wire termination equipment	11%	33%	56%	0%	2	20

NA= number of cases for which data item coded in this is not available. (NA cases excluded from calculation of percentages in table)

Source: Von Hippel (1988, 4)

the majority of innovations relate to intermediary products. The developed technology incorporated in those products is sometimes influenced by the kind and intensity of relationship between the customer and the supplier. Producers of consumer products (final products) buy their components and raw material from suppliers too, but sell their final products in most cases not directly to the consumers, but to retailers or wholesalers. In these situations the relationships between the firms may also influence the technological performances at both sides. Innovation-producing firms may also have special relationships with some of their customers. Von Hippel (1988) studied more than 200 innovations in nine technological fields in detail. He concluded that the majority of innovations were developed by the 'users' of the innovations, not by the manufacturers who bring products from these kinds of technological fields on the market. Table 2.6 offers a summary of Von Hippel's findings.

8   CONCLUSIONS

The current state of knowledge on innovation offers a picture of increasing complexity in the coordination of innovation activities undertaken by individuals and teams from one or a number of organizations. R&D constitutes an important set of activities for a firm to realize innovations, but there are more. The economic literature tends to focus on R&D as the only kind of activity. For the time being we look at R&D as the only set of activities for innovation, but we know that within this set different kinds of activities are undertaken by different actors. By abstracting from the organizational details of the innovation process economists have the opportunity to compare the causes and effects of innovation quantitatively.

At the macro level it is shown that R&D and innovation output have a significant positive effect on economic growth. Although production functions are used as a theoretical foundation, it can be argued that we know that technological change is very important for economic growth, but we do not know enough about how processes and results of innovation at the micro level influence economic growth at the macro level.

At the micro level the impact of innovation on economic growth is less clear. This has to do with the fact that innovations created in one industry may have their main impact on economic growth in one or more other industries. We need to know more about economic theories of innovation at the micro level and their links with macroeconomic growth. In the following chapter the elementary analysis of innovation at the micro level from Chapter 1 is substantiated. No single comprehensive theory is offered, but instead details about the organization of innovation in practice are offered. One theoretical element is discussed in more depth: the relation between producers and users of innovations. This relation is essential for an understanding of how microeconomic processes of innovation lead to macroeconomic growth. Chapters 4 and 5 go into the elements of microeconomic theories of innovation.

NOTES

1 The situation is more complex: new products are used by other firms or by households. In both cases, however, the innovation may result in lower costs of use (see Mansfield et al., 1977).

2 The results were obtained from a count of the patents resulting from the R&D in 1974 of 443 large US corporations.

3 That is to say, a theory which reflects the main elements of innovation strategies and behavior of firms in practice.

4 For a domestic market this assumption would be realistic.

5 Abramovitz (1956), Kendrick (1956), Solow (1957).

6 In some studies the purchase of advanced equipment, such as new CNC machines, is called an innovation. But we will not do this, although the new machine might be new to the firm and some of the problems such as the search for a supplier and the implementation of the machines in the manufacturing process may be similar to some of the problems one is confronted with in innovation activity. The main difference is, however, the creativity element which is absent in the purchase case.

7 It is quite common that a product is improved after it has been introduced in the market. The costs of improvement are known after project completion of the marketed product. One could, of course, treat the improvement activity as a separate project, with its own cost and benefit.

8   We disregard the possibility that a failed project or an invention or prototype not developed as far as commercial application may yield knowledge that is used in other innovation projects. We come back to such complications later.

9   Nelson (1962) gives an early account of the insight that estimations improve during project execution. Many later innovation models still use estimations of cost and benefit ex ante.

10 We mean here after the completion of the innovation project. At the start of an innovation project benefits are more difficult to estimate the more radical is the innovation aimed at.

11 We assume for simplicity that the industry is so disaggregated that we can talk about a market in the economic sense.

12 A few problems remain, such as the appropriate time lag between R&D and commercial results and the number of subsequent years of R&D expenditure which were required to come up with the new products.

13 We saw earlier in this chapter that the majority of innovations are concerned with products used by other firms. Part of them will be parts used in the other firms' products, part will be machines and equipment which may enhance productivity.

14 For a recent study, see Hall (1993).

15 This revealed technology advantage index is calculated in the same way as the relative export performance measure. Note that export performance of country A in industry 1 is not a matter of technological sophistication of the country but of the technological performance of country A's industry 1 compared to the other industries in the country.

16 This report is partly reflected in Brouwer and Kleinknecht (1993b) and in Brouwer, Kleinknecht, and Reijnen (1993).

17 In fact a simultaneous equation model has been estimated in which both the influence of R&D intensity on exports and the influence of exports on R&D intensity has been determined.

18 It is interesting as to whether Dutch firms purchase these machines and equipment from abroad or from domestic firms. In the first case the export performance of the Dutch machine and equipment sectors would decline.

19 Factors such as patents applied for, purchase of licenses, investments in software, office automation did not have a significant influence.

20 They disregard possible differences in the utilization of a technique due to, for example, learning effects or differences in management techniques. Also, only technically efficient techniques are compared: T_i uses more of one input and less of another input than T_j.

21 Many innovations are a mixture of new products and new processes.

22 By a real theoretical foundation is meant here the explication of factors explaining innovation behaviour and outcomes of firms and the relation of innovation results to commercial performance.

23 See, for instance, Pavitt (1982).

24 A remaining problem is the existence of innovative and non-innovative firms in the same industry. Porter argues that not all the firms in a country's exporting industry are necessarily innovative and 'world class' exporters, but the tendency exists that one successful firm stimulates the other and the diamond factors favorite for one are favorite for the others.

25 R&D intensity was included originally, but this variable and some others explanatory variables were highly correlated, causing multicollinearity.

26 It is interesting to note that Nelson (1982, 231) adds as an interpretation for these sectors: 'These tend to preside over rich science bases that can be tapped, often on a custom-tailored basis, to a host of specific industrial demands'.

27 Based on information from patents granted towards firms in that sector.

REFERENCES

Abramovitz, M. (1956), Resource and output trends in the US since 1870, *American Economic Review*, May (Papers and Proceedings), 46, 2, 5–23

Ayers, C.E. (1978), *The Theory of Economic Progress*, Kalamazoo, Michigan, New Issues Press, 3rd edition (first edition 1945)

Brouwer, E. and A. Kleinknecht (1993a), Technologie en de Nederlandse concurrentiepositie; een onderzoek op micro- en meso-niveau, *Beleidsstudies Technologie Economie 23*, The Hague, Ministry of Economic Affairs

Brouwer, E. and A. Kleinknecht (1993b), Technology and a firm's export intensity: the need for adequate innovation measurement, *Konjunktur Politik* 39, 315–25

Brouwer, E., A. Kleinknecht and J.O.N. Reijnen (1993), Employment growth and innovation at the firm level, *Journal of Evolutionary Economics* **3**, 153–9

Griliches, Z. (1994), Productivity, R&D, and the data constraint, *The American Economic Review*, March, 84, 1, 1–23

Hall, B.H. (1993), Industrial research during the 1980s: did the rate of return fall? *Brookings Papers: Microeconomics* 2, 289–343

Hall, B.H. and J. Mairesse (1995), Exploring the relationship between R&D and productivity in French manufacturing firms, *Journal of Econometrics* 65, 263–93

Hippel, E. von (1988), *The Sources of Innovation*, Oxford/New York, Oxford University Press

Ito, K. and V. Pucik (1993), R&D spending, domestic competition, and export performance of Japanese manufacturing firms, *Strategic Management Journal* 14, 61–75

Kendrick, J.W. (1956), *Productivity Trends: Capital and Labour*, New York, National Bureau of Economic Research

Kleinknecht, A. (1996), Innovatie, imitatie en R&D samenwerking: Nederland vergeleken met vijf andere landen, in: F.J.M. Zwetsloot, *De Markt voor Wetenschappelijk Onderzoek*, Utrecht, Lemma, 201–27

Mansfield, E. et al. (1972), *Research and Innovation in the Modern Corporation*, London and Basingstoke, Macmillan

Mansfield, E. (1977), *The Production and Application of New Industrial Technology*, New York, Norton

Nelson, Richard R. (1962), Uncertainty; learning and the economics of parallel research and development efforts, *The Review of Economics and Statistics*, 351–64

Nelson, R.R. (1982), Inter-industry Technology Flows in the United States, *Research Policy* 11, 227–45

Odagiri, H. and H. Iwata (1986), The impact of R&D on productivity increase in Japanese manufacturing companies, *Research Policy* 15, 13–19

OECD (1996), *Technology, Productivity and Job Creation*, Paris

Pavitt, K. (1982), R&D, patenting and innovative activities, *Research Policy* 11, 33–51

Pavitt, K. (1984), Sectoral patterns of technical change: towards a taxonomy and a theory, *Research Policy* 13, 343–73

Rosenberg, N. (1976), *Perspectives on Technology*, Cambridge, Cambridge University Press

Scherer, F.M. (1984), *Innovation and Growth; Schumpeterian Perspectives*, Cambridge (MA), MIT Press

Scherer, F.M. and D. Ross (1990), *Industrial Market Structure and Economic Performance*, third edition, Boston, Houghton Mifflin

Soete, L. (1979), Firm size and inventive activity; The evidence reconsidered, *European Economic Review* 12, 319-340

Soete, L. (1987), The impact of technological innovation on international trade patterns: the evidence reconsidered, *Research Policy* 16, 101–30
Solow, R.M. (1957), Technical change and the aggregate production function, *Review of Economics and Statistics*, August, 39, 3, 312–20
Teece, D.J. (1986), Profiting from technological innovation, *Research Policy* 15, 286–305
Twiss, Brian (1992), *Managing Technological Innovation*, London, Pitman, fourth edition
Veblen, T. (1958), *The Theory of Business Enterprise*, New York, American Library

3. The nature and organization of the innovation process

1 INTRODUCTION

The main goal of this chapter is to provide a general picture of how innovation is managed and organized in practice. To some extent this chapter is complementary to the economics of innovation in the sense that many insights are originating from the management of innovation literature, while the economic approach to innovation tends to abstract from many details of organization. Although I shall give several examples from management of innovation, also in later chapters, it is not the task of this book to provide some sort of synthesis between the economics of innovation and management of innovation. What I intend to do here is to present a 'rich picture' of the economic organization of innovation by the firm which can be linked to several new topics in the economics of innovation literature, several of which are discussed in the second part of the book. In this picture I do not limit myself to the 'internal' organization of the innovation process of the firm. I include the organization of innovation in 'multi-plant' firms and the organization of innovation in industries and even in a whole country. In the latter case one sometimes speaks of 'national system of innovation'. The latter topic is discussed in the last chapter of the book.

The problem

As became clear in the previous chapter, the economics of innovation tend to perceive the innovation processes within the firm as a one-dimensional phenomenon, reflected by the total expenditure of the firm, the total number of man-years devoted to R&D or R&D intensity. Such measures do not account for the differences which may exist between firms with respect to the division of R&D among basic research, applied research, development, and technical services, nor the division among different units, such as corporate laboratories, laboratories devoted to specific technological fields, and divisional R&D departments. As the R&D productivity will differ between these catagories, and the coordination between them and with functional departments of the firm may be organized differently, the link between R&D and commercial performance is affected. Hence, the organization of R&D matters.

A picture is drawn of how innovation is managed and organized in practice. This picture consists of two elements: internal organization and management, and the firm in relation to the technological environment. The description of the organization and management within the firm is largely based on the management literature on innovation. A few concepts from management of innovation are presented here. The picture of the firm in relation to its environment has various sources.

At the end of the chapter some theoretical elements are discussed which touch some of the most important aspects of innovation practice. Some of these elements are based on new economic theories, such as transaction cost economics, but no general framework exists which incorporates all important aspects of the organization of innovation in practice. Within the context of the book it is impossible to systematically link the management concepts from the empirical part to the (rudimentary) theoretical blocks from the economics of innovation literature. It will become apparent that a considerable difference exists between the two.

A first picture

We have already seen that innovators may operate in quite diverse institutional settings and have different strategies, personal characteristics, and represent different organizations. Different actors, different markets and different technologies may very well lead to different organizational solutions for innovation activities.

It has been discussed that innovative activities have been gradually 'institutionalized' in the form of R&D departments of firms. Here we will in the first instance focus on the way in which R&D departments organize their innovation efforts and how the department is embedded in the whole firm. We will go beyond that, however, by introducing new elements of managing and organizing innovation, such as multidisciplinary product development teams, and the management of subcontracting relations. It will also become clear that innovation activity is more than undertaking R&D.

Private inventors and innovating firms are not operating in a vacuum. With respect to the knowledge underlying inventions and innovations sources of technological information can be distinguished. From early on universities have been sources of information for inventors. Later other sources of information occurred, such as public and private research institutes, and today other firms are also sources of technological information; an example of the latter are some main contractors which purposefully provide technological information to their subcontractors. The transfer of technological information from one party to the other is not, or very imperfectly, dealt with through the market. In particular the transfers from public research institutes to small and medium-sized firms is increasingly promoted by a layer of 'intermediary organizations'. Added to this web of organizations can be the various educational institutions which are responsible for the technological, management, and organizational knowledge of newly hired people who undertake invention and innovation activities. Sometimes this whole web of individuals and organizations involved in innovation is called the *innovation system* of a country. Each country may have its unique structure.

Some parts of the innovation system will be relevant for innovation in a specific market, others not. Potentially all firms in a domestic market can profit from the innovation system in the country. Whether they do is dependent on their ability to learn, as we will discuss in Chapter 7, on their geographical location with respect to important technological information sources, on the nature of competition, and so on. Another important aspect is the nature of technology in a market. A simplifying distinction can be made between homogeneous and heterogeneous technology.

When a market is characterized by firms with similar production techniques and similar products (from a technological point of view) we shall speak of a homogeneous market. In such a homogeneous market different organizational solutions for R&D are rather unimportant from an analytical point of view. If such differences exist at all, they may be due to chance or to personal characteristics of people within the firm. This would imply that no systematic analysis of the phenomenon, which relates the organization of R&D to characteristics of the market or the technology in question, is possible. Moreover, it can be questioned to what extent such differences can persist in a strong competition environment. Pursuing this argument, it can be stated that if organizational variety in R&D exists, it is expected to follow from (1) a situation of no or modest competition in the market, or (2) a situation of 'heterogeneity' in production and innovation. That is, the market is inherently characterized by technologically different products and processes. Both conditions are discussed further in Chapter 5 in the context of the nature and extent of competition in innovation. Understanding differences in the kinds of innovation or in R&D productivity between firms in the same market may be understood better when their relations with the innovation system are studied.

It is this condition of heterogeneity which implicitly underlies many of the concepts of management of innovation. One should be aware not to compare them with economic concepts too readily, because these latter may be based on homogeneous market assumptions.

2 THE EMPIRICAL SIDE

Which kind of organizational forms are practiced to undertake innovation? Which management techniques are used? Is the R&D department linked to the other units of the company in various ways? The emphasis in this section

Table 3.1   Inputs and outputs in basic research, applied research, development, and a new product line

Stage	Input			Output	
	(i) Intangible	(ii) Tangible and Human	(iii) Measurable	(iv) Intangible	(v) Measurable
1 'Basic research' (intended output 'formulas')	Scientific knowledge (old stock and output from 1a) Scientific problems and hunches (old stock and output from 1b, 2b, and 3b)	Scientists Technical aides Clerical aides Laboratories Materials, fuel, power	People, hours Payrolls, current and deflated Outlays, current and deflated Outlays per person	a) New scientific knowledge hypotheses and theories b) New scientific problems and hunches c) New practical problems and ideas	Research papers memoranda
2 'inventive work' (Including minor improvements but excluding further development of inventions) (intended output: 'sketches')	Scientific knowledge (old stock and output from 1a) Technology (old stock and output from 2a, 3a) Practical problems and ideas (old stock and output from 1c, 2c, 3c, 4a)	Scientists Non-scientists Inventors Engineers Technical aides Clerical aides Laboratories Materials, fuel, power	People, hours Payrolls, current and deflated Outlays, current and deflated Outlays per person	a) 'Raw inventions' technological recipes patented inventions patentable inventions, not patented but published patentable inventions, neither patented or published non-patentable inventions, published non-patentable inventions, not published minor improvements b) New scientific problems and hunches c) New practical problems and ideas, 'bugs'	Patent applications and patents Technological papers and memoranda Pepers and memoranda
3 'Development work' (intended output: 'blueprints and specifications')	Scientific knowledge (old stock and output form 1a) Technology (old stock and output from 3a) Practical problems and ideas (old stock and output 1c, 2c, 3c, 4a) Raw inventions and im-provements (old stock and output from 2a)	Scientists Engineers Technical aides Clerical aides Laboratories Materials, fuel, power Pilot plants Prototypes	People, hours Payrolls, current and deflated Outlays, current and deflated Outlays per person Investment	a) Developed inventions, blueprints, specifications, samples b) New scientific problems and hunches c) New practical problems and ideas, 'bugs'	Blueprints about specifications new and improved products and processes
4 'New-type plant construction' (intended output: 'new-type plant' and new products)	Developed inventions (output from 3a) Business acumen and market forecasts Financial resources Enterprice (venturing)	Entrepreneurs Managers Financiers and bankers Builders and contractors Engineers Building materials Machines and tools	S investment in new-type plant and products S investment in new-type plant	a) New practical problems and ideas, 'bugs'	New-type plants producing: novel products better products cheaper products process innovations

Source: Freeman and Soete (1997)

lies on the various organizational forms of innovation activity by the firm. The way individual inventors operate, as has been discussed briefly in the previous chapter, has its own influence. Cooperation of private firms with universities and with other firms will be mentioned, but is more extensively discussed in Chapters 8 to 11.

A discription is given of management of innovation and the embeddedness of R&D in the whole company, but we limit ourselves to the essential elements. Management of innovation and technology management are separate fields which the interested reader will have to study separately for a deeper under-standing.

Inventors, R&D departments, and public research institutes

Its has been argued that the role of private inventors diminished in this century and that the role of inventors and teams, operating in R&D departments of private firms became more important instead. Originally universities took care of most of the scientific discoveries and private inventors came up with innova-tions. Today the situation is more complex. Independent inventors still play a role, R&D departments of private firms have become a major provider of innovations, and the importance of public and private research institutes seems to increase.

There is little systematic information available on private inventors. In most industrialized countries associations of private inventors exist and the number of intermediary organizations assisting them seems to increase. Examples of organizations directed towards assistance of individual inventors and operating internationally are the World Invention Trade Association, The British Technology Group, Rayden Ltd and Innovation Partners International (both Japan-based), Licence Marketing International (Germany).

The nature and structure of public and private research institutes differ per country. Public research institutes, like university laboratories, are often fo-cused on basic research in a specific field. Examples are institutes focused on energy, metallurgy, chemistry and transport. When they do applied research it is usually for smaller firms. Private research institutes are normally more directed towards applied research. It is remarkable that almost all develop-ment activity is undertaken by private firms and not by research institutes. The interrelations between R&D departments of private firms and research insti-tutes are in many cases informal and thus not well know. Only when contracts exist between the two do they become publicly known. To a large extent the various organizations involved in innovation have their own place. We discuss this matter further.

R&D inputs and outputs and the process in between

In Chapter 1 a first picture has been drawn of the innovation process and of some major types of innovations which come out of this process. The innovation process consists of a number of more or less subsequent stages of innovation activity, and depending on the kind of innovation output some stage is more important than another. Table 3.1 gives a summary of these stages of the innovation process, and more particularly of the input and output characteristics of each stage. It is based on Freeman and Soete (1997), who refer to Machlup (1962).

Different organizational forms

According to Mowery and Rosenberg (1989) R&D carried out by large corporations has gradually replaced individual entrepreneurs and inventors as the most important source of innovation. We have already described the growing numbers of laboratories (or R&D departments) in the US during this century. The R&D department had remained, until recently, the most common form of organizing a firm's innovation activities. It is responsible for the creation of new products and for preparing new technological solutions in the production process. With the growth and diversification of large firms during this century the need for a large central laboratory concentrating on the technological needs in the longer run increased. The disadvantage of this development was that coordination between product development activity and all sorts of other activity in the firm would decrease.[1] The earlier discussed stepwise model of innovation can be seen as a reflection of such an 'autonomous' R&D department which comes up with a new product, after which coordination with marketing, production, logistics, and so on, takes place.

Table 3.2 Management and organization of innovation processes

Stage	Percentage of total project time	Source/performing group
Idea generation	5%	R&D dept., marketing dept.
Business analysis	10%	Marketing dept., R&D dept.
Technical development Prototype testing	42%	R&D, specific test units
Pre-production	38%	Divisional manufacturer
Market introduction	5%	Divisional marketing dept.

Source: based on Ramsey (1986, 5–8)

The existence of an R&D department as such does not neccesarily mean that the creation and implementation of new technology within the firm takes place according to one set of procedures. In Chapter 2 it has been shown that functionally specialized groups may operate in the R&D department or multi-disciplinary development teams. In the latter, more 'integrated' view on R&D coordination between the R&D department and various other departments of the firm takes place in a more ex ante way.

Ramsey (1986, 5–8) offers an analysis of the management of innovation projects which can be linked to the definition and nature of innovation projects in Chapter 1. He identifies a flow process model of a typical innovation project which is similar to the stepwise model presented earlier. An important distinction is the stage of 'business analysis'. Table 3.2 gives an overview.

Figure 3.1 Four product development team structures

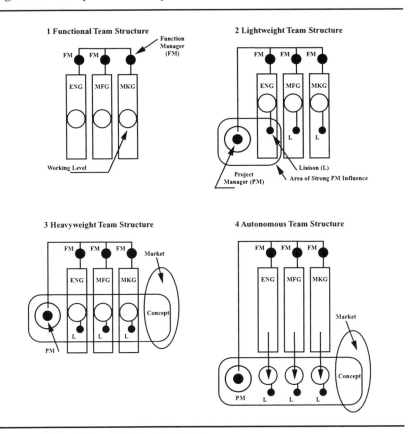

Source: Wheelwright and Clarke (1992, 191)

The process stage model of Ramsey perceives the innovation process as a single system in which the various stages are defined as subsystems. In this sense more interdependence between stages is incorporated than in most stepwise innovation models as reflected in Chapter 1. Nevertheless this picture is 'old-fashioned' in the sense that each stage is assumed to be the responsibility of a functional group or department. Today multi-functional or multidisciplinary product development teams are used.

The links between the R&D department and other (functional) departments, such as production and marketing, can be further illustrated by a discussion of the main types of organization of such modern innovation project teams. Wheelwright and Clarke (1992, 191) make a distinction between four types of structure of innovation project teams: (1). functional team structure; (2). light weight team structure; (3). heavy weight team structure, and (4). autonomous team structure. Autonomous and heavyweight team structures represent crossfunctional teams, while in functional and lightweight team structures people involved in innovation are separated according to functions. Figure 3.1 gives an illustration of the differences between these four team structures.

The four team structures differ in the way the functional tasks are coordinated and in the kind of manager who is responsible for this coordination. In functional teams the various tasks can be assigned to functional experts at the start of the project and each functional manager coordinates the tasks one or more of his people are carrying out.[2] This functional team is actually a loosely coupled set of functional subteams and the main problem is who can be held responsible for the whole product development process, when the original division of tasks among functional department turns out to be problematic. In lightweight teams these coordination problems are reduced by assigning a separate project manager and by assigning a 'liason officer' in each functional department. The people from each department are managed by their own functional manager, who may have more power than the manager who is temporarily assigned to a specific project. According to Wheelwright and Clarke's picture (see Figure 3.1) the project manager will have more influence on functional departments early in the development process (that is engineering) than on departments which become important later on (manufacturing and especially marketing). A heavyweight team is really a crossfunctional team in the sense that the manager of the product development project is 'hiring' people from the different departments to work together and he has the overall responsibility for all work in the project team. This structure enables a better coordination between the different functional tasks within the development project. The autonomous team goes one step further. The project manager is not only responsible for the project outcome, but he may also decide to work differently from the customary way product development is undertaken in the firm. Autonomous teams are normally located separately from the functional departments.

The discussion of various structures of product development teams has shown that which people are indeed carrying out R&D tasks may be unclear in some situations. Particularly in the case of functional and lightweight teams, project members usually devote only part of their time to product development tasks, while their overall performance is evaluated by their functional manager. By contrast people involved in what Wheelwright and Clarke call 'advanced development projects' or 'technology projects' devote all of their time to R&D and these people are located in R&D departments and not in functional departments.[3]

The organizational structure of product development teams is only one aspect of the organization of innovation within the firm. We have already seen that innovation projects other than product development projects exist, such as technology projects. For all kinds of project their existence is something that happens gradually, not suddenly. This means that procedures are needed to come up with project ideas and to select the 'best' projects. In addition a 'technology strategy' is often required, which enables firms to find direction to the search process for new technology.

For the time being we limit our examination of innovation activity in practice to firms where such activity is undertaken on a regular basis. The question is then, how are these institutionalized innovation activities organized and managed, or precisely which institutions are guiding innovative activites? In this respect very large conglomerate firms have a more complex structure of organizing innovation than relatively small firms. We begin with the simple case of a mono-product firm having one or only a few R&D facilities.

Wheelwright and Clarke (1992) are exemplary of the modern view on innovation management. They can serve to illustrate how a firm selects and coordinates its innovation projects (as presented in equations (1.1) and (1.2) and how each of them is managed. A distinction is made between individual project plans and the overall innovation strategy plus the aggregate project plan in which individual projects are selected, coordinated and evaluated. But first a remark. In comparison with what has been said about innovation in Chapters 1 and 2 (and with some of the analysis of the economics of innovation which is yet to come), Wheelwright and Clarke see a strict separation between R&D in general and concrete plans to develop specific products or processes in practice. In what they call advanced development projects, firms are undertaking research for new product or process technologies and in the development of these new technologies the project teams need to arrive at a stage of 'proven technology'. Only proven technologies are (together with the already used set of technologies in the operational divisions/departments) considered as part of a specific product development project (Wheelwright and Clarke, 1992, chapter 2). The argument for such a strict distinction is interesting. They use 'invention' for the development of such new technology and development project for the application of one or more new technologies in new products or processes for commercial development:

> When invention (for which the timing, prerequisites, resources, and specific out-comes are largely unpredictable) is included in a development project, it invariably causes delay, backtracking, and disappointment. However, when done in advance so that its results are available for application, development of new technology may contribute significantly to project success. The implication is that required inven-tions should be proven (i.e. feasibility demonstrated) beforehand, off the critical path of commercial development. (ibid. 38)[4]

Thus, from a planning perspective the stages of research and development of new technology should be separated from development of (and thus applica-tion of new technology in) commercial new products and processes.[5] Implicitly, many economic models of innovation are apparently mixing technology strat-egy (in which it is decided which new technologies are required for the firm's competitive position in the next years and in which some new technologies are developed and tested until 'proven') and product development plans.[6] The question remains to what extent technology can be 'proven' without commer-cial application in the form of product applications of the new technology.

We make a distinction between an innovation project, the organization of a set of projects related to a specific product market (multi-project planning) and the overall organization of innovation in the firm. The overall organiza-tion relates to the technology and innovation strategy chosen and the location of various research and product development facilities. Multiproject planning concerns procedures used to plan, select and monitor different projects. Pro-ject planning refers to the specific way in which innovation is carried out in the firm.[7]

The overall organization of innovation activity of the firm

Ideally a firm has a technology strategy and this strategy fits to an overall strategy. In this technology strategy a vision is put down on how new technol-ogy will be used to reach the firm's mid- and long-term goals. Included may be a patent and licensing strategy, a strategy on technological cooperation, and an innovation strategy. This innovation strategy should include a formulation of the kind of technologies the firm wants to create in the coming years, and the ways in which and places where it can be used. It could also include the way innovation is organized in terms of central laboratory, divisional R&D departments, product development teams, and so on. According to Ramsey (1986) firms often lack a clear procedure of how to integrate the technology strategy process with the overall strategy process.

In large firms basic research, applied research, and development activities – the basic distinction introduced in Chapter 1 – are often carried out in different kinds of innovation units and at different locations. Moreover some of the development activities are carried out by people working in functional departments such as engineering, production, and marketing.

The globalization of R&D

The situation of R&D in multiproduct firms has been briefly discussed. One of the main questions is to what extent, in addition to economies of scale in R&D, the specific form of organization (that is, coordination between the various R&D units) of innovation in the multi-R&D-unit firm influences innovation performance. Today an increasing number of firms operate as multinational enterprises (MNEs) and although production and distribution facilities tend to be spread more over the world than R&D units, these MNEs are also confronted with a number of R&D units in different countries. Florida (1997) states that especially in the US many MNEs have started R&D facilities to 'harnass external scientific and technological capabilities and generate new technological assets' (ibid., 85). These foreign 'subsidiaries' accounted for more than 15 percent of total industrial R&D expenditure in the US in 1994. Westney (1990, 283) distinguishes four types of foreign R&D units of the MNE:

1. Technology transfer units (TTUs): to facilitate the transfer of the parent's technology to the subsidiary, and to provide local technical services.
2. Indigenous technology units (ITUs): to develop new products for the local market, drawing on local technology.
3. Global technology units (GTUs): to develop new products and processes for world markets.
4. Corporate technology units (CTUs): to generate basic technology for use by the corporate parent.

It is clear that communication between all the units of the various types is required to establish an effective and efficient innovation process within the MNE. We come back to these coordination problems in the theoretical section.

Sources of innovation other than R&D units

There may be other sources of innovation than the various innovation units mentioned in the foregoing. This is apparent in small firms which quite often do not have an official R&D department (see below). More generally engineering, software development, and (aesthetic) design are activities which are not included in the firm's R&D activity, but which do contribute to the number and quality of innovations a firm produces.

Table 3.3 Relationships of R&D strategy to stage of industry development

Stage of industry development	Emphasis of R&D strategy			
	Offensive	Defensive		Licensing
	New products New technology	Product improvement	Process improvement	
Phase 1				
Rapid growth	High	Low	Low	Low
Low competiton				
Phase 2				
Market growth	Medium	High	Medium	High
Increasing comp.				
Phase 3 Maturity				Medium
Low growth	Low	Medium	High	to
High competition				high

Source: Twiss (1992, 81)

The engineering department controls the firm's production processes. In this role engineers may come up with improvements of existing machines or with ideas for more fundamental process innovations which they will usually transfer to the R&D department. In addition they may also make suggestions to improve the existing product(s) design in such a way that large scale production is less costly or can be realized at higher quality standards.

In between 'multi-project planning for a single market' and 'technology strategy' lies the organization of innovation and technology for a number of markets. Portfolio analysis is an important tool from this area of management which is of interest to economists. The aim of portfolio analysis is to assess the economic potential of each market the firm operates in and the firm's position in each market. Quite often markets are classified according to the stage of the industry's life cycle. Table 3.3 illustrates this.

Next to the kind of R&D strategy required in each stage is the position of the firm in each stage and the related strategy. Figure 3.2 shows this.

Multiproject management

While the technology strategy is directed at what the firm wants, multi-project management is concerned with how the innovation objectives can be reached. Questions dealt with are 'how can we generate enough ideas?', 'how can we bring these ideas into successful innovations?', and 'how can we select between various projects?'.

Figure 3.2 R&D strategy in a multi-product firm

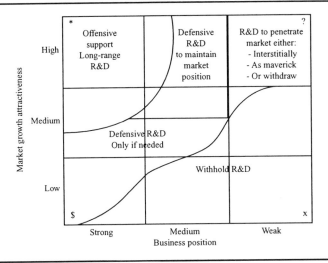

Source: Frohman and Bitondo (1981), in: Twiss (1992, 82)

In some cases firms will have less money for R&D than they have ideas for new technology development and new process and product projects. In such cases it is obvious that one is in need of selection criteria to decide which projects will be carried out and which not. But even in the case one expects not to spend more money than is available when all projects are carried out, the 'rational economic spirit' demands that criteria are used to decide how much to spend on each project.

Shepherd (1990) is an example of a rather restricted and abstract economic way of selection of projects. He argues that one should list the projects according to the expected 'return on investment' (ROI.) and spend money on each additional project as long as this ROI is positive and above a certain minimum level, to be set by the firm. Figure 3.3 is reflecting these ideas. Note that this 'economic' allocation of resources to alternative projects results in an 'optimal R&D intensity' of the firm, which will ususally fluctuate quite strongly over the years. An alternative approach is to allocate resources to projects with an ROI above the mentioned minimum until a fixed budget is exhausted (indicated by the vertical line in Figure 3.3).

It is not to be denied that managers must carry out project selection procedures in which expected benefits and costs of each project play an important role, but there may be some complications in practice. First, costs and benefit estimations may be unreliable and they often change during project execution. This could mean that some projects will be cancelled after a period of time.

Figure 3.3 Cost and benefits of R&D projects

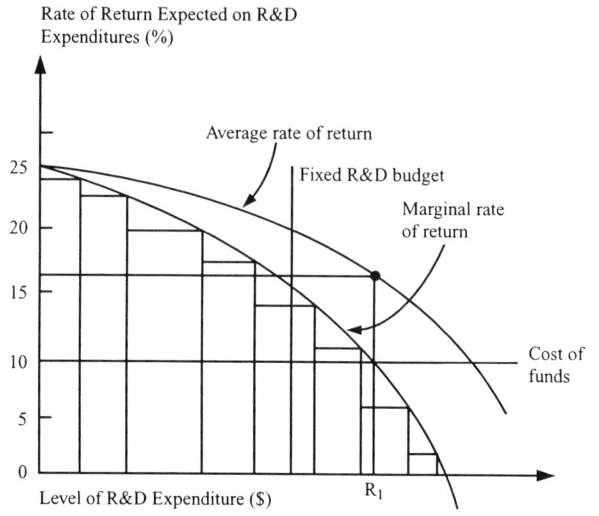

Rate of Return Expected on R&D
Expenditures (%)

Average rate of return

Fixed R&D budget

Marginal rate
of return

25

20

15

10

Cost of
funds

5

0

Level of R&D Expenditure ($) R_1

Source: Shepherd (1990, 147)

Second, we have seen that there may be a great difference between new product development projects and projects aiming at the development of an entirely new technology for the firm. It is very difficult to estimate the r.o.i. for a radical development project, because the new technology will not be used in new product development before it is proven. Third, and related to the second point, innovation projects may be of a different kind. For example, the current market position of a firm may be decisive as to whether to carry out a radical innovation project, a 'new platform' project,[8] or an improvement project for existing products. A technology strategy is therefore an important frame of reference in addition to the estimated costs and benefits of individual projects.

Project management

When it is decided to start a project, firms often follow a set of procedures for projects of a certain kind. One such procedure could be to select a specific type of product development team, as depicted in Figure 3.1. Other procedures have to do with control and coordination. One such procedure is to set up a time schedule for a product development project in which it is planned when the project uses specific test facilities, and how much time each researcher is devoting in each stage (this is especially relevant when more than one project is carried out simultaneously). An example of coordination is the so-called

Table 3.4 Innovation activity outside the R&D department in the Italian manufacturing industry

Selected industries	Number of firms with innovations from 1981–85	Percentage of firms with R&D expenditure in 1985
Metals	148	31.8
Chemicals	490	70.8
Metal products	1051	20.0
Mechanical engineering	1404	44.7
Office machinery[a]	11	90.9
Electric, electronic	658	58.7
Motor vehicles, parts	177	49.2
Food	319	17.6
Textiles	714	13.6
Footwear, clothing	412	9.5
All industries[b]	8220	31.1

[a] Computing products.
[b] The study contained more industries than depicted here.

Source: Napolitano (1991)

'early-supplier involvement', where in an early stage of the product development process the team discusses with one or more suppliers which kind of new parts may be needed in the near future.

The position of small firms

The organization of innovation in the firm is first of all dependent on the size of the firm. Many small- to medium-sized firms do not have an R&D department, but a number of them do carry out innovation activities. Table 3.4 shows some figures for Italy of innovating firms without any offically registered R&D. In such smaller firms innovation activity may be limited to design and engineering, or may be limited to part-time work of employees whose main tasks lie elsewhere in the firm.

Up to here to picture has been drawn that the R&D of an innovating firm or other organization is the single source for inventions and innovations. This picture is incomplete. Individual inventors, project teams, R&D departments, and research institutes depend to some, or even a large, extent on external information and previous technological achievements to create innovations.

Each innovating unit is part of one or more systems of innovation from which they draw information.

The national system of innovation

According to Mowery and Rosenberg (1989) technological developments and applications by private firms in practice have as much initiated and directed basic research in the scientific community as scientific development dictated subsequent R&D by private firms (the last causation is suggested by the step-wise model).

For each country a picture can be drawn that gives an overview of the most important categories of organization involved with technological innovation. One type of organization may be an information source for an innovating firm. Such a source may be another organization and even a competitor. We come back to the information sources of innovating firms later in the book. For now it suffices to know that each country may have its own peculiar 'technological infrastructure' or 'system of innovation'. We come back to the working of the technological infrastructure in Chapter 11.

The outsourcing of innovation/R&D

In the last decade many industries have shown a development where firms sell parts of their production. They sell lines of business which are outside the 'core competencies' and outsource activities to suppliers in their major lines of business. This model of 'lean production' is supposed to increase efficiency and to better follow the technological developments in the narrower remaining field(s) of technological competence. Related to this is the need to innovate faster and with less R&D budget. This led quite recently to outsourcing of parts of R&D as well.

One of the fields where outsourcing of parts of the R&D has been quite common is the pharmaceutical industry. This had to do with the typical nature of pharmaceutical research that new 'prototype medicines' have to be tested on animals and humans. Large hospitals are better equipped to do this than R&D departments of pharmaceutical companies. Another example from the same industry comes closer to the current need for outsourcing of R&D: large pharmaceutical firms source out (part of) their biotechnological research to small specialized biotechnology firms.

Outsourcing of a part of the firm's R&D activity implies a rather well-defined set of tasks which is separable from the other tasks, carried out by the firm itself. The importance of such R&D contracts for large European firms has been increasing in the early 1990s.[9] According to Hagedoorn and Schakenraad (1992) outsourcing is an important form of technological cooperation. Note that R&D contracts can be held with other firms, especially suppliers, as well as public research institutes. A consequence of the outsourcing of parts of R&D is that the firm becomes more dependent on R&D activities undertaken outside the firm.

3 SOME THEORETICAL ASPECTS

As is the case with organization in general, the organization of innovation is not discussed very much in economics literature. What are the theoretical underpinnings of the various forms of organizing innovation? We limit the discussion largely to the internal organization of innovation, but outsourcing of R&D activities is included in the analysis. Various forms of technological cooperation are explained in Chapters 8 to 10.

A basic distinction can be made between creativity and technological problem solving as such – activities which are undertaken in R&D departments, not exclusively but predominantly, and coordinating these activities with numerous other activities in the firm in order to make innovations (including implementation) as effective and efficient as possible. Creativity is usually assigned to the inventor while the coordination problem is one of the tasks of the innovator or entrepreneur. We have seen that Schumpeter made a sharp distinction between entrepeneur and inventor. In his opinion the entrepreneur has the more important role, because by making new combinations (Neue Kombinationen) of existing technology he could even do without inventions.

The economics literature has hardly dealt with creativity. We will discuss some theoretical elements of individual creativity and teamwork below.

Coordination problems have much more been of concern to economists. 'Transaction cost economics' (TCE) and 'agency theory' are two of the most important theories of economic organization. Largely, however, they have not been used to explain the prevailing forms of organization of innovation. In empirical research, however, a considerable number of studies has drawn on transaction cost economics in formulating hypotheses on the organization of innovation. Therefore, TCE in relation to innovation will be discussed.

The creativity concept of Usher

Although autonomously operating inventors are not as important as in the early days the creativity of individuals is still very important. Concepts exist that try to establish the right conditions in the firm, but creativity ultimately depends on the characteristics of the people involved. Usher (1962, 58–69) makes a distinction between three approaches that explain how the creative element in invention occurs. The 'transcendentalist' approach attributes the emergence of invention to the inspiration of the occasional genius who from time to time achieves insight into essential truth through the exercise of personal energy, intuition, and skill. Invention, and thus innovation, is dependent on the presence of such individuals. This implies that the time and place of inventions are difficult to predict. This picture comes close to the old inventors and not to firms that try to plan the process and timing of innovations more and more. The 'mechanistic' process (the second approach) views invention as proceeding under the stress of necessity with the individual inventor being an instrument of historical processes; inventions are produced continuously. Individuals are dependent on the accumulation of knowledge on a certain domain, so firms specialized into such a domain will sooner or later create inventions. Diffusion of knowledge is important in this approach and

Figure 3.4 A creativity concept

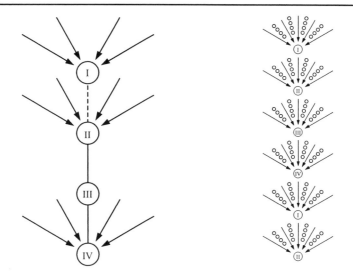

The emergence of novelty in the act of insight: synthesis of familiar items:
I.   perception of a incomplete pattern;
II. the setting of the stage;
III. the act of insight;
IV. critical revision and full mastery of the new pattern.

The process of cumulative synthesis. A full cycle of strategic invention, and part of a second cycle. Figures I–IV represent steps in the development of a strategic invention. Small circles represent individual elements of novelty. Arrows represent familiar elements included in the new synthesis.

Source: Usher (1962, 67–9)

one firm could improve its inventive record by managing the stream of public knowledge. While the transcendentalist approach emphasizes the individual, the mechanistic approach underscores the prominence of the scientific community where one idea is following from many previous ones. Tapping from the 'common pool of knowledge' is strongly influenced by the institutional setting of the inventor. The third approach, the 'cumulative synthesis approach' combines the core elements of the others. Major inventions are visualized as emerging from the cumulative synthesis of relatively simple inventions, each of which requires an act of insight. Figure 3.4 depicts this combination. In practice it means that many small inventions can be realized or used by the firm by creating the proper institutional setting. One point is that the inventor must have access to the common pool of knowledge. This opens up the possibility for relatively small firms to invent without being dependent on the one genius. Major inventions are still following from the insight of one or a few individuals and are thus unpredictable.

Usher focuses strongly on the cognitive aspects of invention, although the institutional setting of the inventor implicitly recognizes other aspects, such as the organizational culture that stimulates creativity. As we have seen, investments in the stages from exploring ideas until the 'prototype' are relatively inexpensive. Developing the prototype on to a commercially successful product is more costly, in general less uncertain and less dependent on creativity. One can therefore say that Usher overemphasizes invention (Scherer, 1984). As Jewkes et al. (1969) pointed out, many inventions from individuals or small firms are not carried through to innovations, or when they are, large firms are involved. To a certain extent technology projects that are kept out of planning procedures with strict time and money budgets (see our discussion of Wheelwright and Clarke at the beginning of this chapter) are dependent much more on inventors than product development projects. In the latter the 'entrepreneur' who sees the commercial opportunities of a combination of (existing) technologies, organizational procedures and the like is more important. The inventive aspects of product development are more and more depending on teamwork and less on a single genius. One can argue, however, that the organizational talent of the entrepreneur is as rare and unpredictable as the 'technological' talent of the inventor. It is not clear whether the concepts of Usher can be used for the organizational genius as well.

Cross-functional coordination

In the empirical section different team structures have been discussed in relation to the coordination of the main functional activities: production, marketing, and engineering. As Wheelwright and Clarke (1992) make a distinction between invention (research) and development projects (development and engineering), one may add R&D to the list of main functional activities of the modern firm. It has been shown how team structure might explain the nature and extent of cross-functional coordination (integration, according to Wheelwright and Clarke, 1992). Of course several other factors explain cross-functional co-ordination in practice. The point here is that firms may differ in the way they coordinate their R&D with the other activities, which no doubt influences R&D productivity and the overall competitiveness of firms. The same holds for the organization and coordination of innovation in various innovation units.

Coordination among R&D units

Central in the coordination among innovation activities and of innovation with other firm activities is the question of centralization–decentralization of R&D. Coombs and Richards (1993) provide a historical account which shows that R&D in large firms used to be strongly centralized and has been decentralized in the 1980s. The advantages of centralization are, among other things, coordination of

Table 3.5 Main features of the autonomous and linkage models

Autonomous model	Linkage model
Existed since World War II	Implemented in the mid-1980s
Research funds generated as a result of a flat tax on the company's business divisions	research funds generated as a result of direct contracts from the company's business divisions
Scientists generate projects on the basis of company's generic interest	Scientists and managers generate projects on the basis of customers' needs
Corporate R&D enjoys autonomy from the rest of the company	Corporate R&D depends on the rest of the company
Emphasis on long-term research	Emphasis on short-term research
Technical feasibility directs the research	Availability of money directs the research
Strategies for R&D are clear	Strategies for R&D are vague
Indirect links between research and business divisions	Direct links between research and business divisions
Emphasis on research	Emphasis on development
Scientists work on few and similar projects	Scientists work on many and different projects
More layers of management	Fewer layers of management

Source: Varma (1995, 236)

the various innovation projects (which also avoids duplication), economies of scale and scope, and a climate of creativity. The advantages of decentralization are that innovation activities are more focused on the needs of the operational units and the speed with which they are completed is higher. Varma (1995) discusses this matter in more detail in terms of 'autonomous' model and 'linkage' model (decentralized). Table 3.5 provides the main features of both models.

The core idea for our purpose is that innovation performance is influenced by the degree of centralization and by the organization of R&D in more general terms. This organization, as the distinction of four types of R&D units of MNEs as Westney has already suggested, has as its main objective to combine 'all' the technological opportunities of the various units with the local market opportunities. In this respect it is of interest to emphasize Florida's conclusion from a survey of 186 foreign firms' R&D units in the US that technological motives played a slightly larger role than market motives for beginning an R&D subsidiary in the US (Florida, 1997). We will see with the discussion of technological cooperation, later in the book, a similar distinction between market motives and technological motives.

Transaction cost theory and outsourcing of R&D

Transaction cost economics (TCE) is concerned with the organization of economic activity at two levels: at the level of individual transactions and at the level of the multi-product firm.[10] At the transaction level the theory addresses the conditions under which specific products or services are organized within the firm (internal delivery) or with an outside supplier. As the R&D function can be seen as a service to the primary activities of the firm, such as production, the same question can be posed whether a specific innovation project is carried out by the firm itself or by an outside research unit.

At the multi-product level TCE focuses on the coordination of the various operational and strategic activites of the firm. With regard to innovation this could include questions about the location of a R&D department, for example under which circumstances a central laboratory is the best solution in a multi-divisonal firm and when separate divisional R&D departments are the most efficient. The question here is similar to the centralization question of Coombs and Richards. The underlying motives are different, however. TCE stresses the importance of the attenuation of opportunistic behaviour in cases where the two parties in a transaction undertake specific investments. We will pursue this explanation further in the case of outsourcing of R&D. In the following the principles of TCE are briefly discussed. Thereafter we focus more specifically on the explanation of outsourcing of R&D.

TCE assumes that economic actors are boundedly rational and behave opportunistically. Bounded rationality means that actors intend to take rational decisions, but they are hampered in doing so by the limited amount of information they have on each decision topic. Limited information stems from the complexity of a problem, the changes that occur over time, and the fact that some of the other actors do not provide information, or when they do, it may be incomplete or misleading. The latter has to do with opportunistic behavior. When an economic actor wants to maximize his profits he sometimes purposefully withholds the information on which the partner in an economic transaction (for example a supply firm) can optimize his decisions. But of course it could also be the other way around. All actors may behave egoistically and the problem is that they try to hide their intentions to maximize their own profits at the cost of the trading partner. By closely monitoring the other actor a firm tries to reduce the chance of opportunism. But monitoring is costly: it causes *transaction costs* to rise. As a result some transactions are deliberately executed in-house. So when a firm anticipates that a supplier may behave opportunistically and it is difficult to neutralize the effects, it decides to buy the supplier or to set up an internal supply unit.

The products and services that a firm needs for its own product(s) vary in complexity, in price, in quality, in volume, in the nature of technologies involved, and so on. According to Williamson (1985) three characteristics of the transaction of the product or service are decisive for the kind of governance structure the firm creates to attenuate opportunism: asset specificity, frequency, and uncertainty. Asset specificity is high when investments made by a party in a transaction cannot be used for the transaction of another product or service, or can be used only at high 'transformation' costs. An example is a machine that can only be used to produce a component specific to a single customer. When asset specificity is high the parties in the trading relationship become dependent on each other and this makes one vulnerable for the consequences of opportunism. As a result transactions with high asset specificity tend to be internalized by the firm that needs the product or service in question. The frequency of the transaction is the number of times the product or service needs to be transacted, say per year. When the frequency of the transaction is low and when the set-up costs are relatively high internalization may be uneconomical. Uncertainty is defined as the difficulty to predict the other parties, behavior and the difficulty to understand the strategies and developments in the 'supply' market more generally. Consequently information about the market and information provided by a (potential) supplier is difficult to assess. Hence opportunism is not recognized quickly. A high degree of asset specificity, frequency, and uncertainty leads to internalization or 'unilateral' governance of the transaction (one party is in command). Figure 3.5 reflects the different organizational forms chosen under different combinations of the three characteristics of the transaction.

Figure 3.5 Cost efficient organizational solutions for the various combinations of transaction characteristics

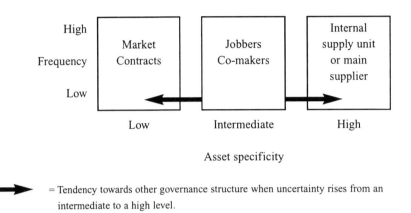

= Tendency towards other governance structure when uncertainty rises from an intermediate to a high level.

Source: based on Williamson (1985)

In practice it regularly occurs that a firm develops a component and tests the product line or process with which to produce the product. Then it specifies the technical characteristics of both the product and the manufacturing process and orders a supplier to produce and deliver the product according to the specifications. Under these circumstances the firm is perfectly able to evaluate the information provided by the supply firm about price, delivery schemes and other variables in the draft contract. For innovation projects the situation may be entirely different. By definition the technical specifications of a new product cannot be formulated in advance. Consequently the outsourcing of part of the development tasks can only be considered at a very late stage of product development, for instance after extensive testing of the prototype and after pilot production. When the market dictates rapid product development one loses valuable time with such a late involvement of suppliers. The alternative is early supplier involvement without knowing exactly what is needed. This would be a special case of relational contracting. Relational contracting occurs when internalization is not economical and when the danger of opportunism is high. Both parties invest specifically in the relationship to signal 'good intentions'. An example is a supplier who finances product development for a specific customer itself in anticipation of a contract by a customer. In return one or more researchers from the supplier are being informed about the product development process of the customer, including the plans for future product generations in which the supplier could play a role. Early supplier involvement leaves one vulnerable to opportunism, because investments are by definition specific. One of the advantages is that the customer can use the technological and market expertise of the supplier in its own product development process.

The link between the organization of innovation and performance

TCE assumes that when outsourcing of R&D is chosen, it is done so because the production costs and transaction costs of R&D are thus minimized. More generally one could argue which theoretical arguments give us insight into the linkages between the organization of innovation and performance of the firm. We have already discussed the complexities to establish a link between R&D inputs and innovation outputs. In economics there are very few theories concerned with organization of innovation and performance. Agency theory is especially applicable at the level of teams but has not been applied to innovation at this level of the firm.

The structure–conduct–performance approach in economics deals with the organization of innovation and performance at the industry level. This issue is discussed in Chapter 5. There we also discuss Henderson and Clarke (1990) who link organizational features of firms with their ability to react to new product introductions by competitors.

Here we limit the discussion to the study of Bean (1995). He has linked his research on organizational factors of R&D to total factor productivity (TFP) as discussed in Chapter 2. The growth of TFP over time (ΔTPF) is measured in terms of differences in growth rates of outputs and inputs. It has been shown that 'traditional' production factors only account for about 30 percent of TFP. The remainder is due to unexplained factors, of which changes in the stock of technical knowledge is argued to be the main influence. When traditional input factors labor, capital, and material are held constant, and when R&D intensity is assumed to reflect accumulation of the stock of knowledge, we get:

$$\Delta TFP = a + b(R\&D \ intensity) \tag{3.1}$$

On the basis of estimations of R&D intensity and annual TPF growth rate of 15 US drug companies between 1971 and 1990 it was found that a = 0.10 and b = 0.62. From previous research it is known that the part of R&D devoted to product and process development contributes significantly to TFP. In addition Bean (1995, 26) found that the greater the level of basic and applied research in the R&D program, the greater the contribution of product and process development to annual TFP growth and the greater the level of TFP growth. Also the emphasis on external sources of technical knowledge in the 1980s positively influenced TFP growth. In contrast the complexity of the R&D organizational structure was negatively correlated with TFP. In particular, when business group/sectoral laboratories were found above corporate and divisonal laboratories, TFP growth turned out to be lower. This obviously has to do with coordination between R&D units (see section on (de)centralization of R&D).[11]

Relationships between creators and users of innovations

The simple picture that can be drawn on the basis of the technology flow schemes presented in Chapter 2 is that of industries where all firms are undertaking R&D, and other industries where the firms are not performing any innovation activity, and they are merely purchasing the new technology from the first category of industry. Reality is far more complex; the economic world cannot be divided neatly into creators and users of innovation. In this section we first discuss a taxonomy of industries of Pavitt (1984) based on various mixtures of innovation and the use of new technology. Thereafter we examine the work of von Hippel (see, for example von Hippel, 1988) who studied the relationships between innovator and user firms in practice and came up with a few theoretical concepts.

The input and output indicators of new technology stress (implicitly) the importance of the creation of innovations. When applied properly, they measure the efficiency of firms, industries, and whole countries in 'producing' new technology. To the extent that suppliers and customers influence the innovativeness of firms, the measurement of R&D productivity by exclusively looking at the inputs and outputs of the innovating firm may sometimes be misleading. For an understanding of the innovativeness of the firm the R&D intensity of its suppliers could be important too.

Different types of industry

In Chapter 2 it became clear that industries differ with respect to the number of innovations used and produced by their firms. In agriculture, for example, most innovations used there are created by suppliers of agricultural equipment. By contrast most innovations which are applied in the microelectronics industry are also created there.

Pavitt (1984) analyzed empirical data for the UK about approximately 2000 significant[12] innovations from 1945 to 1979 and came up with a classification of industries according to a number of factors determining the nature and pattern of technology creation and use. Four types of industry are identified: 'science-based', 'supplier dominated', and 'production-intensive' which is divided into 'scale-intensive', and 'specialised suppliers'. The names indicate the main source of innovation of that type of industry: science, suppliers, and the producers in the industry themselves.

Apart from the significant innovations themselves, the database contains information about the industry in which the innovation has been created, the industry in which it is used, and the principal activity of the innovating firm.[13] Another important feature of the innovating firm is its size. Other main categories in the database are the nature of the innovation (product or process innovation) and the sources of the innovation, in particular sources within the firm and sources outside (for example another firm or a university). In 58.6 percent of the cases intra-firm sources are mentioned as the most important source for the innovation. This leads Pavitt (1984, 348) to conclude that public knowledge as a main source of innovation (as neoclassical theory assumed according to the stepwise model) is not what practice shows:

> The concept of the general 'pool' or 'stock' of knowledge misses an essential feature of industrial technology, namely, the firm-specific and differentiated nature of most of the expenditures producing it ... Whilst it may be reasonable to describe *research* and *invention* as producing 'information' that is quickly and easily transmitted, it is grossly misleading to assume that *development* and *innovation* have similar properties. [emphasis in the original]

But now back to the main aspect of the study, an understanding of the patterns of production and use of innovations. In the 'brief history of technology in economic science' in Chapter 2 it became clear that the creation of technology was largely seen as exogenous to economic processes. In other words, industries are mainly users of innovation which is produced outside (that is in the scientific community). This appeared to be largely true for agriculture, the service sectors,[14] and manufacturing industries such as textiles. These sectors are called 'supplier dominated'. For most manufacturing industries, however, the story is different. Table 3.6 gives a summary of Pavitt's findings. It shows a classification of sectors at the two-digit level according to the four types of 'technology sectors' and a number of underlying characteristics of each type of industry.

It must be stressed that the classes are ideal types; an industry may have many characteristics of one of the industry types, but in general combinations occur. The automobile industry, for example, can be seen as a mixture of scale-intensive and supplier–dominated. Another way to look at Pavitt's classification of industries is to see whether product or process innovations dominate and who the creators and users of these innovations are. Product innovations dominate in two types of industries. In the one (science-based) the firms are large (on average), they innovate mainly within the sector (limited diversification) and few outsiders come up with new products. In the other type (specialized suppliers) firms are usually small, many innovations are created by firms outside the industry (especially large customers such as automobile

Table 3.6 Sectoral technological trajectories: determinants, directions and measured characteristics

Category of firm	Typical core sectors	Determinants of technological trajectories			Technological trajectories	Measured characteristics			
		Source of technology	Type of user	Means of appropriation		Source of process technology	Relative balance between product and process innovation	Relative size of innovating firms	Intensity and direction of technological diversification
1	2	3	4	5	6	7	8	9	10
Supplier dominated	Agriculture; housing; private services; traditional manufacture	Suppliers Research extension services; big users	Price sensitive	Non-technical (e.g. trademarks, marketing, advertising, aesthetic design)	Cost-cutting	Suppliers	Process	Small	Low vertical
Production intensive — Scale intensive	Bulk materials (steel, glass); assembly (consumer durables & autos)	PE suppliers; R&D	Price sensitive	Process secrecy and know-how; technical lags; patents; dynamic learning economies	Cost-cutting (product design)	In-house; suppliers	Process	Large	High vertical
Production intensive — Specialised suppliers	Machinery; instruments	Design and development users	Performance sensitive	design know-how; knowledge of users; patents	Product design	In-house; customers	Product	Small	Low concentric
Science based	Electronics/ electrical chemicals	R&D Public science; PE	Mixed	R&D know-how; patents; process secrecy and know-how; dynamic learning economies	Mixed	In-house; suppliers	Mixed	Large	Low vertical / High concentric

PE = Production Engineering Department

Source: Pavitt (1984, 354)

manufacturers) and the innovations that the firms create are aimed at other industries (technological diversification). Examples are mechanical and electrical engineering. In the other group mainly process innovations occur. In the supplier-dominated sectors these innovations are largely created by suppliers. In the scale-intensive type of industries most process innovations occur by the firms in the industry and are innovations for that industry.

User initiated innovation

In his explanation of technology flows between sectors Pavitt (1984) stresses the patterns of production and use of innovations, thereby suggesting that firms in an industry either create innovations or use innovations which have been created in another industry. He also argued, however, that in many industries other firms are important sources for the innovations created by firms in the industry. One of the examples mentioned is that manufacturers in the automobile industry create a considerable number of process innovations, which would normally have been created by the machine industry. Here the link can be made with the innovations created or initiated by suppliers or customers of manufacturers in a specific industry (von Hippel, 1988). We wil come back to these and other forms of technological cooperation.

4 CONCLUSIONS

Economists sometimes reduce the whole innovation process to a one-dimensional picture: the level of R&D expenditure of the firm. The empirical part of this chapter revealed many more dimensions at various levels of analysis (project, firm strategy and organization, interfirm relationships). The theoretical part made clear that no uniform economic concept of organization of innovation exists. In fact economic theory has largely ignored this aspect. In several of the following chapters we come back to the institutional and organizational aspect of innovation.

NOTES

1. See Mowery and Rosenberg (1989) for an account of this development in the US
2. In the examples given here functional specialists are originating from the large traditional departments such as R&D, production and marketing. In practice it is also possible that a technology unit or R&D department is divided into subgroups such as electrical engineering, mechanical engineering, process engineering and so on.
3. These advanced development projects are discussed below.
4. A question not answered directly by Wheelwright and Clarke is how new technologies can be proven without any application in concrete new products or processes.

5. A disadvantage of such a sharp distinction is that it suggests that invention in the sense of creating new technological knowledge only occurs in advanced development projects and not in 'normal' development plans. New technological findings are made in both the stages until prototype realization and the subsequent stages of further development and testing. This implies that a patent can also be obtained in all stages (see Chapter 7).
6. Of course, the overall innovation strategy and organization includes standard procedures to be followed by (various kinds of) projects.
7. Of course, the overall innovation strategy and organization includes standard procedures to be followed by (various kinds of) projects.
8. A platform project is defined by Wheelwright and Clarke (1992) as a project aiming at major product innovation that opens up the possibility for a series of new products, based on the major technological change.
9. The increase of outsourced R&D in large European firms is documented by a number of consultancy reports from Sweden, Germany, Switzerland, France, the Netherlands and the USA carried out for the Dutch Ministry of Economics.
10. The distinction is not always fully recognized. Williamson (1985) emphasizes the study of single transactions being any transaction between two technically separable stages of production (or, more general as any transaction where an implicit or explicit contract is involved). The transaction cost framework is directed at this level. Transaction cost considerations are also kept at the level of the multi-product firm, such as the 'Multi-divisional firm'. The original 'logic' of transaction features matched by certain organizational solutions is not completely kept, however.
11. Other factors not dicussed futher in this book are: the technology planning horizon (+), the expectations of technical managers about R&D contributions to business objectives (+), and the perception of government regulatory requirements as a factor in R&D project selection (-).
12. Significant innovations were identified by experts knowledgeable about, but independent from, the innovating firms; information about the characteristics of the innovations was collected directly from the innovating firms (Pavitt, 1984, p.344).
13. This last element has to do with a technical matter. It is possible that an innovating firm in industry X has created a type of innovation which is 'normally' created by firms in industry Y. This means that the innovation falls outside the firm's principal activity (which is located in industry X). It is possible that the firm will diversify its production into industry Y. Likewise the innovation may be used in industry Z, but also by a firm which has its principal activity outside industry Z.
14. Software development is not defined as 'innovation' here.

REFERENCES

Bean, A.S. (1995), Why some R&D organizations are more productive than others, *Research Technology Management*, January–February, 25–29

Coombs, Rod and Albert Richards (1993), Strategic control of technology in diversified companies with decentralized R&D, *Technology Analysis and Strategic Management* 5, 4, 385–96

Florida, R. (1997) The globalization of R&D: results of a survey of foreign-affiliated R&D laboratories in the USA, *Research Policy* 26, 85–103

Hagedoorn, J. and J. Schakenraad (1992), Strategic partnering and technological cooperation, in B. Dankbaar, J. Groenewegen and H. Schenk (eds), *Perspectives in Industrial Organization*, 171–91, Dordrecht, Kluwer

Henderson, R. and K. Clarke (1990), Architectural innovation: the reconfiguration of existing product technologies and the failure of established firms, *Administrative Science Quarterly* 35, 9–30

Hippel, E. von (1988), *The Sources of Innovation*, New York/Oxford, Oxford University Press

Jewkes, J., D. Sawer and R. Stillerman (1969), *The Sources of Invention*, London, Macmillan

Machlup, Fritz (1962), *The Production and Distribution of Knowledge in the United States*, New Jersey, Princeton University Press

Mowery, D.C. and N. Rosenberg (1989), *Technology and the Pursuit of Economic Growth*, Cambridge (MA), Cambridge University Press

Napolitano, Giovanni (1991), Industrial research and sources of innovation: a cross-industry analysis of Italian manufacturing firms, *Research Policy* 20, 171–78

Pavitt, Keith (1984), Sectoral patterns of technical change: towards a taxonomy and a theory, *Research Policy* 13, 343–73

Ramsey, Jackson Eugene (1986), *Research and Development*, Michigan, Ann Arbor, UMI Research Press, University Microfilms

Scherer, F.M. (1984), *Innovation and Growth: Schumpeterian Perspectives*, Cambridge (MA), MIT Press

Shepherd William, G. (1990), *The Economics of Industrial Organization*, Englewood Cliffs, Prentice-Hall

Twiss, Brian C. (1992), *Managing Technological Innovation*, fourth edition, London, Pitman

Usher, Abbeycot P. (1962), *A History of Mechanial Inventions*, Cambridge (MA), Harvard University Press, second edition

Varma, R. (1995), Restructuring corporate R&D: from an autonomous to a linkage model, *Technology Analysis and Strategy Management* 7, 2, 231–47

Westney, E. (1990), Internal and external linkages in the MNC: the case of R&D subsidairies in Japan, in: C. Bartlett, Y. Doz and G. Hedlund (eds), *Managing the Global Firm*, London, Routledge

Williamson, O.E. (1985), *The Economic Institutions of Capitalism*, New York, The Free Press

Wheelwright, S. and K. Clarke (1992), Revolutionizing product development, New York, Free press

4. Main problems at the micro level

1 INTRODUCTION

In the previous chapters it has become apparant that both innovation and the use of new technology are important from an economic point of view. In this chapter we refocus on the innovators and thus on the creation process. What are the main problems of creating innovations? The chapter is divided into two parts. First, empirical studies are discussed about the failure and success of innovation and the uncertainty of innovation projects at the micro level. Second, the theoretical side is covered with the appropriability problem as the core.

The Problem

The outcome of an innovation project is uncertain by definition. The so-called innovation paradox makes this clear: If you know in advance the information (knowledge) that comes out of an innovation project you already have what you were looking for, so you do not need to start the project. It will become clear that the information that is meant here is largely technological information. The uncertainty with respect to this kind of information is called 'technological uncertainty'. Projects differ with respect to the number and complexity of the technological problems to be solved. As a result the degree of uncertainty of the projects differ. In projects aiming at an improvement of an existing product, for example, some or a large part of the information is known (or at least perceived as known) and one can concentrate on the 'small' part to improve. Sometimes it turns out that during the improvement process many new problems arise, so sometimes technological uncertainty appears to be greater than anticipated. For other projects, such as technological or advanced development projects, uncertainty is very high.

Why do firms invest so much in innovations, despite this inherent uncertianty? In the introductory chapter we already mentioned that competition is a strong driving force to innovation. Being forced to innovate is one thing, being successful is another matter. What are the main problems innovators encounter? An important distinction must be made between the firm level and the market level of analysis. At the market level competition might force firms to invest in innovation projects, despite the uncertainty they are confronted with. At the firm level the uncertainty of innovation may lead to managerial and organizational problems. For example, banks are reluctant to finance a firm's innovation projects. Consequently even firms with a very good track record in innovation might still have problems in acquiring external funds for R&D. In this chapter we concentrate on problems at the firm level. In Chapter 5 technology competition at the market level is the core. Both topics are, however, strongly intertwined.

In Chapter 3 it was argued that an effective and efficient management of innovation can enhance the speed of innovation, improve the quality of innovations and at the same time reduce innovation costs. It is important to understand that the focus of 'the microeconomics of innovation' discussed in the current chapter is different. Microeconomics is concerned more with the interaction between firm and market than with the internal organization of innovation in each of the firms. One could say that management of innovation is taken as an exogenous factor and that other (endogenous) factors are examined. In addition the economics of innovation tends to look for patterns of innovation behavior or innovation results of groups of firms rather than the preferred interaction of a particular firm and its market(s). This partly explains why the economics of innovation comes up, for example, with empirical results showing a decrease in R&D productivity instead of an organizational mode to increase this productivity. Other differences come up, because economics tends to treat R&D and innovation as homogeneous categories. This may conceal differences between creativity and 'mere' development of prototypes. At the same time, however, the economics of innovation is concerned with aspects of creativity, such as uncertainty of innovation projects, whereas management of innovation tries to exclude invention (radical development projects, Wheelwright and Clarke, 1992) from its field of interest, because it is too uncertain.

The uncertainty of innovation projects is revealed most clearly in the number of projects that are started but that do not result in commercial success. In the economics literature a distinction is made between technological uncertainty and market uncertainty. Technological uncertainty relates to the specific technology and the internal organization of the firm. Market uncertainty relates to what competitors of the innovating firm are doing and the theoretical analysis at this point focuses on the appropriability problem. This last mentioned topic is very prominent in the literature. Mowery and Rosenburg (1989) state that the emphasis neoclassical economics puts on the appropriability problem has caused the organization of R&D to be neglected. It shows yet again that economics is not focused at the firm, but at the market: the degree in which innovation profits can be appropriated by the firms on average. The two topics can nevertheless be linked. The relation between the organization of R&D of the firm and its ability to appropriate benefits from its innovations is treated by Teece (1986).

2 EMPIRICAL EVIDENCE ON INNOVATION PROBLEMS

Problems innovating firms may have are no doubt often a combination of problems following from competition in their market(s) and problems inherent to the technology they focus on, to personal and organizational limitations, and so on. As said before we focus on other than market problems here.

Firm-specific problems

Creating new technology requires a considerable level of technological expertise which can only be acquired through a mixture of theory (education, know-how) and practice (experience from actually undertaking innovation activities) over a number of years. Even then coming up with a prototype and testing it is to a certain extent a matter of trial and error.[1] After the test phase one still does not know whether one succeeds; the market (users) must ultimately decide. What are the main problems of firms in coming up with successful innovations? Let us as usual first examine empirical studies on the topic.

Ramsey (1986) gives an overview of empirical studies concerning project failures. In his opinion the largest of such studies is of Booz, Allen and Hamilton[2] and they state that only two percent of the project ideas that enter the new product development process stage model are introduced successfully into the market place, and that about three-fourths of the funds used in the process are used on projects that are not successful. Ramsey (ibid., 41) furher states that according to this study:

> this general success and failure average differs surprisingly little among widely varying industries. Since the state and rate of change of technology differ widely among industries, this identification of the relative constancy of success rate suggests that technical considerations and the state of the technical art in that industry do not play a major role in determining the new product success rate of a firm.

That does not mean that the inability to solve all kinds of technological problems is not causing projects to fail, but the chance that a project fails is not dependent on the technology or industry the firm operates in.

In the so-called SAPPHO project 29 pairs of similar innovation projects in the UK during 1966–72 were studied, of which the one project failed and the other was a commercial success. As each pair of projects was carried out in the same industry, the success and failure factors for these 58 innovation projects could be established irrespective of the industry/technology. The study identified the following five factors as being markedly different for successful and unsuccessful projects:

- Understanding of user needs;
- Attention paid to the marketing of the new product;
- The efficiency of the development work, but not necessarily the speed;
- The effectiveness of the use of outside knowledge and advice (but not as a substitute for in-house work);
- Seniority and authority of the individuals in the project.

All these factors are related to the organization of innovation and the characteristics of the people involved. Remarkably this study, and most comparable studies (see, for example, Twiss, 1992), say nothing, or at least remain implicit, about technological uncertainty. Admittedly it may not be the main cause of project failure, and uncertainty generally is difficult to handle, but it is quite often a topic in more theoretical discussions on innovation at the firm level. A few empirical studies have asked explicitly about technological uncertainty. In the theoretical section we discuss the classification of types of innovation according to technological plus market uncertainty (Freeman, 1982).

Another matter is to what extent problems of innovating firms are related to the kind of innovations they are aiming at. The SAPPHO study focused on radical innovations. Others have not made a distinction between different types of innovation. It seems obvious that the more radical an innovation is the more it is away from the firm's core technological competence accumulated up to that moment, and the more the firm has to explore unknown areas. Some innovations involve simply a more complex technology than other innovations or even a whole aray of technologies. And although technology does not explain failure or success among industries, radical and complex innovations are expected to have a higher level of technological uncertainty and thus a higher failure rate than incremental innovations in the same industry.

Mansfield et al. (1972) provide results of a study on the innovation process of thirteen major chemical firms and eight major petroleum firms in 1964. Part of the research was about technological uncertainty of innovation projects. Table 4.1 gives data on the probability of technical success as perceived by the managers in the 21 firms. A project is defined as being a technical success if it attains its technical objectives in the budgeted time and within the budgeted cost (ibid., footnote 4, 20). Technological uncertainty can subsequently be defined as the uncertainty as to whether a project results in new or improved products or processes, and in addition whether it does so within the planned time and budget.[3]

From Table 4.1 it follows that the technological uncertainty of the firms' innovation projects is not perceived as very high. Twelve out of the sixteen firms for which figures are available perceive that more than half of their innovation projects have at least a 50 percent chance of technical success. Other studies show that the actual rate of technical success of innovation pro-

Table 4.1 Percentage distribution of projects by estimated probability of technical success

Chemical firms probability of technical success	1	2	3	4	5	6	7	8	9	10	11	12	13
0–24%	80	na	0	10	na	36	20	25	15	0	20	15	5
25–49%	10	na	0	20	na	18	20	25	15	50	20	35	20
50–74%	0	na	0	40	na	18	30	25	30	50	20	35	60
75–100%	0	na	80	20	na	18	10	25	20	0	40	15	15
not known	10	na	20	10	na	10	20	0	20	0	0	0	0

Petroleum firms	1	2	3	4	5	6	7	8
0–24%	11	na	na	25	0	na	10	5
25–49%	11	na	na	5	10	na	5	10
50–74%	22	na	na	10	10	na	5	30
75–100%	36	na	na	50	30	na	60	40
not known	20	na	na	10	50	na	20	15

Source: derived from tables 2.1 and 2.2 in Mansfield et al. (1972, 22–3).

jects is not much different, although considerable variation among firms and industries may exist. As mentioned before, Ramsey (1986) concludes that the majority of causes of innovation project failure are non-technical in nature. An important question related to this technological uncertainty is that, although it is not perceived as high and not much variation seems to occur among firms, the time and cost devoted to a project will influence the perceived technological uncertainty. We come back to this point in the theoretical section.

The most important question perhaps is why Ramsey, Twiss and others conclude that only a small minority of all projects are successful, while Mansfield states that the technological uncertainty of innovation projects is in most case not very high (half of the total number of projects had a probability of more than 50 percent to be completed as planned). One explanation could be that the 'status' of innovation projects in the Mansfield study and the Ramsey study differs. Mansfield focused on existing innovation projects (that is, product development projects), while the Booz, Allen and Hamilton study referred to by Ramsey included all initial project ideas that entered the firms' selection process. Another thing is that the definition of success is different. Mansfield uses technical completion, while Ramsey uses commercial success.

Commercial success is not only dependent on (a reduction of) technological uncertainty, but also on (a reduction of) market uncertainty. Especially in practical situations which approximate 'a market of homogeneous technology' firms will have similar innovation projects. The chance that a competitor has a successful project could make another firm's project a failure. The question arises: why is one firm doing better than another, despite the fact that the other firm is also following the rules of management with respect to conditions for success (for example the five SAPPHO factors)? Three explanations can immediately be given: first, a competitor started earlier with its innovation project and completed the project earlier (with an equal quality of innovation management and human capital); second, the competitor had on average a better score with regard to the success factors; third, it had more luck in solving problems in the various stages of product development.

But even when the firm markets a new product first and is the innovator in the market, many cases show that the innovator loses and a competitor, a customer, or a supplier wins most of the profits stemming from the original innovation. Teece (1986) gives examples of innovating firms which win or lose (see Figure 4.1)

In the SAPPHO study and other studies managerial and organizational factors turned out to play an important role in predicting the success or failure of an innovation project. The economic approach in terms of market uncer-

Figure 4.1 The firms which won most of the profits from an innovation

	Innovator	Follower-imitator
Win	1 • Pilkington (float Glass) • G.D. Searle (NutraSweet) • Dupont (Teflon)	2 • IBM (Personal Computer) • Mutsushita (VHS video) • Seiko (Quartz watch)
Lose	4 • RC Cola (diet cola) • EMI (scanner) • Bowmar (pocket calculator) • Xerox (office computer) • DeHavilland (Comet)	3 • Kodak (instant photography) • Northrup (F20) • DEC (personal computer)

Source: Teece (1986, 287)

tainty and technological uncertainty tends to disregard such matters. There are exceptions. In Link (1993, 8–9) a study is discussed in which information related to the evaluation of the return on R&D was collected from 88 manufacturing firms. Each R&D vice president was asked if their unit annually evaluated, formally or informally, the outcomes of their R&D efforts by estimating a return to their R&D investments. Not surprisingly, all 88 responded that they did. Of these 88, 36 reported that they utilized a formal evaluation method and 52 indicated that their method of evaluation was informal and often varied from year to year. In general, the methods considered to be formal in nature are based on an analysis of ratios. Those methods described as informal were varied, ranging from expected sales or cost savings per R&D project to the percentage of R&D projects completed within the budget. No single evaluation method was consistently mentioned by any of the vice presidents.

The study by Link (1993) also concluded that for the firms using a formal method this evaluation was considered by the R&D vice presidents to have a major impact on the R&D budget (a mean response of 2.98, where 3 = major impact, 2 = minor impact, and 1 = no impact), while the score for firms using informal methods was 2.25. In addition, for the 88 firms, a positive correlation was found between size of the R&D activity (R&D/sales) and the choice for a formal evaluation method and between formal evaluation method and firm performance (using 'total factor productivity growth').

Industry-specific problems

Market competition and the market uncertainty following from it is the topic of the next chapter. Industries or markets do, however, differ in another respect which may influence innovation processes of the firms considerably. This additional factor is called technological opportunities. Technological opportunities can be defined as the opportunities for innovation each firm in an industry has due to the scientific developments related to the industry's technologies and the technological developments in the industry as a whole, which become public knowledge to each individual firm. The greater the technological opportunities in an industry the more each firm may choose a technological subfield which best fits the current knowledge base of the firm. Related to this is the possibility of finding a niche market. It is difficult in empirical research to measure differences in technological opportunities. Table 4.2 indicates the significance of technological opportunities with respect to four output indicators.

Technological opportunities are used in empirical analyses where the impact of competition (concentration, firm size) on the innovativeness of the firms in different industries is measured, to correct for differences in autonomous technological development. Apart from the fact that technological opportunities are difficult to measure it is questioned today by some authors

Table 4.2 Technological opportunities and industries

	Percentage of residual variance explained			
	Patents	**Gross revenues**	**Profits**	**Market value**
Technological cluster effects				
Alone	21.2*	23.0*	19.4*	28.5*
With industry effects	14.1*	7.8*	4.1*	12.3*
Industry effects				
Alone	8.5	28.4*	24.2*	30.4*
With technological effects	2.4	8.1	6.5	15.2*

* Statistically significant at 99% level.

Source: Jaffe (1989, 93)

whether such differences still exist. Despite the popular distinction in high-tech, medium-tech, and low-tech industries, some authors argue that new technological developments are pervasive in every industry. Every competitive firm is deeply involved in innovation and the application of numerous technologies, no matter the industry they are in. And indeed some state that there are high-tech and low-tech firms in each industry.

One could still argue, however, that one technology is causing more problems than another and that industries differ with respect to the complexity of the technologies with which they have to compete.

This specific nature of technology is often disregarded by economists. In the theoretical section general features of technology are discussed, such as tacit and codified knowledge, but apart from such general classifications for all technologies, other technology-specific problems might play an influential role.

3 THE THEORETICAL SIDE

The theoretical explorations of innovation problems at the micro level tend to focus on the effect of the market on the individual firm. The terms 'market uncertainty' and 'technological uncertainty' originate from this kind of literature (Kamien and Schwartz, 1982). In such studies innovation is mostly dealt with as if it were a 'homogeneous' phenomenon; it does not matter what kind of innovation is aimed for and the activity undertaken to realize innovation is the same everywhere: R&D. Sometimes research is distinguished from development. We first follow this literature and then deal with the influence of different types of innovation.

Market uncertainty and technological uncertainty

Each innovation project is subject to market uncertainty and technological uncertainty. Market uncertainty is defined as the uncertainty of the innovator with respect to its profits from innovation due to the actions of competitors and customers. Originally market uncertainty was distinguished from profit uncertainty. (Kamien and Schwartz, ibid.)[4] Here we take the two together, because they are both concerned with characteristics of the market the innovator is operating in. Technological uncertainty is concerned with problems which have to be solved while carrying out R&D (or innovative activity more in general). As discussed previously this technological uncertainty causes difficulty in estimating the costs and/or the completion time of an innovation project. Below we focus on technological uncertainty. Market uncertainty is related to the appropriability problem which is dealt with in the next subsection.

Technological uncertainty exists because people carrying out an innovation project will encounter technical and technological problems they are not yet familiar with, and that they can thus not solve immediately. Only by successfully solving some technological problem, or combining untried combinations of existing technological solutions, does one come up with a technological innovation. The creativity process, analyzed by Usher, can be seen as an important step in solving such problems.[5] Undertaking R&D can thus be seen as gradually solving 'all' the problems, and by doing so more and more knowledge is accumulated. In the beginning of a planned project the technological uncertainty is the highest and with the accumulation of R&D effort this uncertainty is slowly reduced. Figure 4.2 gives a stylized version of this process of reduction of technological uncertainty.

Level I can be seen as the knowledge required to solve all the technological problems. Level II could be the level at which additional problems occuring after market introduction (or in-house use) are solved. This knowledge accumulation curve assumes diminishing returns to scale as is often the case in economics. One could well argue that, at least during some periods of the project, increasing returns occur, because insight to solve one problem may be the breakthrough to solve a number of related problems. Curve A in Figure 4.3 shows an alternative pattern of knowledge accumulation, with increasing returns to scale in the middle stage. An additional complexity to the knowledge accumulation process is that knowledge may accumulate due to R&D, but this knowledge may not be relevant to the current project. It may even hap-

Figure 4.2 Knowledge accumulation and uncertainty reduction

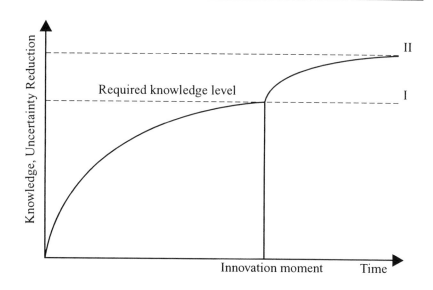

pen that one is gaining insight that a specific problem cannot be solved with the existing stock of knowledge in society and the project must be cancelled, which of course is an essential element of technological uncertainty.

We have seen that technological uncertainty is not perceived by managers as a major cause for project failure and that it does not vary among different industries. This implies that the probability that the accumulation of technological knowledge reaches the desired level for technological solution is relatively high, and that it merely depends on the size of R&D expenditure (and as we will see later on, the time path of expenditure). And indeed, in many mathematical models of innovation it is assumed that the rate of technological uncertainty is only dependent on the level of R&D expenditure during a specific time period. It must be realized, however, that the number of empirical studies in which probabilities of technical completion of projects are estimated is still very low. Innovation theory has to address two problems yet: (1). differences in the level of technological uncertainty between types of innovation projects; (2). An increase of uncertainty across projects over time.

In Chapter 3 it was stressed that the product development process must be separated from technology projects, because the latter type of innovation projects were too uncertain to be managed. More generally it can be argued that various types of projects are expected to have different levels of technological uncertainty. According to Mansfield et al. (1972, 41) a clear positive correla-

Figure 4.3 An alternative pattern of knowledge accumulation

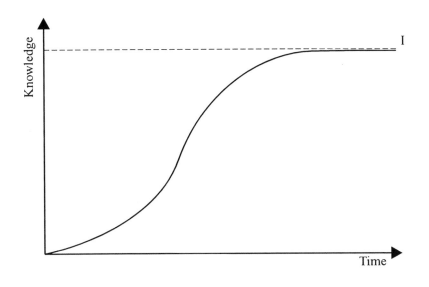

tion exists between the probability of technical completion and various indicators of project innovation types.

During the last decades a tendency has existed for the technological uncertainty of a typical innovation project to increase. This is due to, amongst other things, increased international competition between firms in most markets, which forces innovators to explore more unknown technological areas and speed up the innovation process. Another reason is the increased number of technological fields, which must be combined to create innovations. Speed and complexity of innovation projects are therefore the keywords behind technological uncertainty. Increasing the speed with which to complete an innovation project, or choosing a more complex project, will both tend to increase the total R&D costs of the project. This is consistent with the knowledge production function in the sense that higher speed and more complexity require more knowledge to be produced per time period, hence higher R&D expenditure per time period.

The relation between the speed of innovation and total project costs is known as the time–cost tradeoff curve shown in Figure 4.4. This time–cost tradeoff curve shows that the total R&D costs of an innovation project increase when the time of completion of the project decreases. The completion time is the period from the beginning of the project until the successful ending of the project, that is, when the new product is introduced in the market or when a new process is applied in-house. Originally the curve reflects the cost and lead

Table 4.3 Technological uncertainty in different kinds of innovation projects, based on over 200 projects in three laboratories

Laboratory and type of product	Probability of technical completion
Laboratory	
X	0.68
Y	0.66
Z	0.52
Extent of technical advance sought	
small	0.66
medium	0.42
large	0.26
Type of project	
product improvement	0.72
new product	0.49
Familiarity with technical area	
familiar	0.59
unfamiliar	0.46
All projects	0.57

Source: derived from (Mansfield et al., 1972, 41, table 2.10)

time of a large number of different innovation projects. The projects with the shortest lead time tend to have the highest cost. Several explanations are given for this phenomenon (Scherer, 1984, 67). One is the basic assumption that diminishing returns set in as more work force is applied to individual project tasks. Another is related to technological uncertainty.

Diminishing returns per unit of labor cost could be due to increased labor cost per hour or to diminishing returns per unit of labour cost per se. The first explanation is that R&D employees need to do overtime work, which in most countries is more expensive than regular wage costs.[6] The more important reasons concern diminishing returns, for instance because too much pressure to complete a project may result in failures or in not enough time to learn from internal or external sources about existing solutions to encountered problems. Both cause a rise in project costs, because extra man-hours, materials, and

Figure 4.4 The time–cost tradeoff curve of a typical innovation project

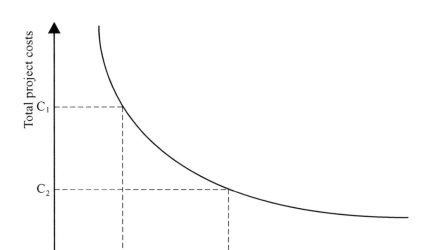

equipment are required to repair failures or to find solutions, which are a waste because without knowing these problems have already been solved by others outside the R&D-project team. This last point implies that the knowledge production function as presented in Figure 4.2 is not only dependent on the volume of R&D and the way this spending is managed, but also on the effective use of external sources. This aspect of utilization of external sources becomes quite important when firms innovate while at the same time they do not carry out R&D.

At a more fundamental level Scherer (1984, 68) relates the convexity of the curve to the number of approaches a firm can follow to solve a technological problem. The intuitive notion behind this is that the higher the number of parallel approaches the higher the probability of success, but also the higher the cost of innovation. We dig into this problem a bit deeper, because the idea of parallel approaches plays an important role at the industry level too.

Let N be the number of approaches. Each approach is assumed to have the same prior subjective probability of success p, where $0 < p < 1$. Consequently, $q = 1\text{-}p$ is the probability of failure for a given approach. We further assume that the success probability p of an approach and its cost M are independent of the number and sequence of other approaches pursued by the firm. Finally, success of one approach means that the technological problem is solved (the innovation project is completed) and the other approaches are not needed any-

more. Under these assumptions, the probability of overall project success is
defined as:

$$P(p,N) = 1 - q^N \qquad\qquad (4.1)$$

It is clear that the probability of project success increases with the number
of approaches taken, but obviously the total project cost will increase too.
Both depend on how the N approaches are phased over time. 'One possibility
is to run the approaches serially, that is, one after the other. Another is to run
all the approaches concurrently in the first period, guaranteeing project completion by the end of that period Scherer (1984, 69).[7] An interesting conclusion
of Scherer (ibid., 71–2) is that,

> Under these assumptions the most efficient strategy is to schedule a greater num-
> ber of approaches in each succesive time period ... In common sense terms this
> means that when interested in achieving success by the end of two or more periods,
> one saves the bulk of the trials for later periods, hoping that the few early trials will
> yield a success making the expense of later approaches unnecessary.

Two remarks about the assumptions above: first, learning between approaches does not alter the conclusion of convexity of the function, it merely
shifts the curve to the left for T > 1 (ibid., 75).[7] When the N approaches do not
give exactly the same solution, things become different (ibid., 76): 'In this
respect minimum-time strategies and series strategies cannot be completely
comparable, since the concurrent execution of many or all approaches will
yield more successes than will series execution of the same approaches, with
project termination after the first success'. Second, another important conclusion
is that 'the greater the payoff in future time periods contingent upon successful research project completion, the more concurrent rather than series
scheduling of research approaches should be emphasized' (ibid., 77).

Mansfield et al. (1972, chapter 7) provide an additional explanation for the
increase of total project costs when innovation processes are speeded up. First
of all they distinguish the minimum expected time and cost of an innovation
project. These are the boundaries within which the time–cost tradeoff curve
lies. Table 4.4 provides estimates of the minimum expected cost and time of
29 innovation projects.

Mansfield et al. (1972, 136) assume that the time–cost tradeoff function
can be represented by a relatively simple equation, the parameters of which
vary from innovation to innovation:

$$C = \frac{\Phi}{v\,e^{t/\alpha-1}} \qquad\qquad (4.2)$$

Table 4.4 Estimates[a] of minimum expected time and cost for 29 innovations

Innovation	Cost[b]	Time[c]	Innovation	Cost	Time
1	225	32	16	544	24
2	84	24	17	185	15
3	22	8	18	201	12
4	274	15	19	9155	60
5	133	12	20	5220	21
6	163	9	21	2100	120
7	52	6	22	1982	48
8	2610	18	23	1140	60
9	15	5	24	620	48
10	209	24	25	160	24
11	276	18	26	1082	60
12	32	18	27	1403	60
13	354	20	28	595	34
14	437	24	29	753	23
15	292	20			

[a] The values of R^2 (adjusted for degrees of freedom) vary for the 29 innovations from 0.61 to 0.98, with a mean of 0.83.
[b] The cost are expressed in thousands of dollars.
[c] The completion time is expressed in months.

Source: Based on table 7.1 of Mansfield (1968, 138)

Where C is the expected cost of innovation, t is the expected time, α is the minimum expected cost, v the minimum expected time, and Φ an unknown parameter. Φ, v, and α vary from innovation to innovation.

A basic characteristic of the time–cost tradeoff function as represented here is that the closer the expected time of a project is already to the expected minimum possible, the higher is the increase in total project cost for an additional speed-up of the project. At the same time there seems to be a considerable variation among projects (different values of Φ). Mansfield identifies various determinants of the elasticity of cost with respect to time. We discuss the most important ones. First, he distinguishes the complexity of the project. The more the innovation advances the state of the art the higher is the value of the elasticity. For such projects Mansfield (ibid., 141) argues that 'there is obviously greater initial uncertainty and more learning to be done than in less ambitious projects. Later stages of the project depend more on what has been learned in earlier stages than they do in less ambitious projects. Thus, it tends to be more expensive to speed up more ambitious projects.' The size of the firm is a second variable. To the extent that time reduction requires a certain flexibility of approach, large firms may be handicapped by inertia and by more difficult administrative problems (longer chain of command, and so on). Related to this can be the size of the project. 'If large projects require more coordination and integration of different tasks, they may be relatively more complicated and costly to speed up.' (ibid., 142). So if large firms have on average larger projects the two factors reinforce each other.[8]

The third factor mentioned is previous experience in the field of the innovation. This factor relates to the previously mentioned interrelatedness of a firm's innovation projects over time. Mansfield perceives two opposing effects. On the one hand researchers and engineers with previous experience may have become a specialist in solving problems typical for the field. On the other hand (considerable) time reductions may require completely new procedures and problem solving methods, and thus people with different mental models.

Complexity and size of the innovation project, and its fit with previous projects, all have to do with technological uncertainty. If we assume that each task of an innovation project has a certain level of technological uncertainty[9] and we define the size of a project as the number of tasks T and complexity as the number of relations R between these tasks, it follows that the 'overall' uncertainty of a project increases with complexity and size. The influence of experience could in addition be defined in terms of the number of tasks which are common to tasks which have been solved successfully in previous innovation projects.

In the tradeoff between time and cost the easy solution would be to avoid high costs, that is, not to speed up one's project(s) and not to choose complex projects, and so on. However, competition in the market may force a firm to speed up its innovation process and to come up with more complex innovations. We will see this in the next section where market uncertainty is discussed.

Mansfield et al. (1972, 145–52) give an interesting analysis of the influence of the overlap of different stages of the innovation process on completion time. So, if firms are forced to speed up their innovation processes, it may be that they look for an overlap of stages as an organizational solution. But this overlap tends to increase the total cost of an innovation project and, hence, provides an explanation for the time–cost tradeoff function. Recall that the stepwise innovation model of figure 1.4 contains a number of subsequent stages. Table 4.5 presents figures on the frequency of overlap between five stages of the 29 innovation projects which we used before.[10] Two forms of overlap are distinguished. 'Pattern 1' concerns two stages of which the shortes stage starts later, but ends at the same time as the longer stage. In 'pattern 2' stage i starts and ends earlier than stage i + 1, but there is a considerable overlap.[11]

Table 4.5. Frequency of occurrence of overlap between stages of the innovation process

Overlap Between

Type of Innovation and Pattern of Overlap	stages 1 and 2	stages 2 and 3	stages 3 and 4	stages 4 and 5
	(number of innovations)			
Chemical innovations				
Total number	6	10	8	6
Pattern 1	5	0	1	6
Pattern 2	0	4	4	0
Other	1	6	3	0
Nonchemical innovations				
Total number	5	2	10	10
Pattern 1	3	0	0	7
Pattern 2	0	0	7	2
Other	2	2	3	1

Source: Table 7.6 in Mansfield et al. (1972, 147)

Innovations have different total completion times and stages may have a different length too. If, however, the length of the time over which stages overlap is related to the length of these stages, we get a relative measure on which innovations could be compared. Equation (4.3) gives a definition of a so-called overlap structure of an innovation project.

$$V_i = \frac{l_i}{t_i + t_{i+1}} \tag{4.3}$$

where the overlap structure is a four-element vector (V1, V2, V3, V4).

In addition to the speed of the innovation project the size of the firm may give an explanation for the value of V for each innovation project. The argument is that some stages are undertaken by different units and that more generally big firms have more specialized workers to enable overlap. The degree of overlap between each pair of stages indeed seems to correlate with the speed of innovation and with firm size, and there are indications that a high degree of overlap could explain high cost.[12]

The discussion of technological uncertainty has brought an interesting problem to the surface. While many modern studies on the management of innovation stress the need to speed up the innovation process and to lower innovation costs simultaneously, some economists argue that speeding up will imply that innovation costs rise. It may be concluded that speeding up the innovation process has its limits. Up to a certain point organizational slack may be reduced and costs decreased accordingly, but after this point additional speeding up will lead to an increase of innovation costs.

In many innovation models the increase of R&D gives diminishing results, but an increase of knowledge as such remains possible. Implicitly a high if not perfect mobility of R&D labor is assumed, which reduces the problem of knowledge accumulation to an appropriate allocation of people with required levels and fields of expertise. In practice, however, R&D activities are firm-specific and newly hired researchers and engineers need considerable time to learn the specific – often tacit – know-how of R&D, production, marketing, and so on, before they can contribute to the innovation process. In addition it can be argued that no one inside the firm is able to know precisely which people to allocate to which innovation project, as such information would require knowledge which is yet to follow from the results of the projects for which people have to be hired. An optimal allocation of resources towards innovation projects as illustrated in Chapter 1 thus not only requires reliable estimations of costs and benefits, but also insight about specific expertise ex ante.

In contrast with most neoclassical models of innovation, evolutionary economics stresses the firm-specific nature of R&D. Related to this assumption it puts more emphasis on rules of thumb to decide about the aforementioned matters about which one by definition does not have accurate information. This could induce firms to allocate resources not to all possible projects with high expected net profit, but to favorable projects close to previous projects.

Evolutionary economics

Evolutionary economics tries to understand the interaction between firms in a market and the development of the economic system as a whole. In this respect the approach is similar to neoclassical economics. It differs with respect to the behavioral assumptions of the firm and with respect to the position of technology. In analogy with evolution theory, selection and mutation play an important role. Mutation is strongly related to new products and processes, but organizational change is not neglected. Nelson and Winter (1982) see innovation as a change of routines, of habits and explicit rules and procedures to perform tasks. Selection of new products (directly) and new processes (indirectly) is steered by the market; the customers decide which new products they prefer. Mutations do not arise at random. Firms purposefully follow a certain direction in their search. In the chapter on learning it will be discussed that people and firms undertake innovation projects which are often 'as close as possible' to the existing core of technological expertise. This is what evolutionary economists call 'path dependency'. On the one hand there is deliberate search and path dependency. On the other hand there is luck and surprise. For example, the invention of nylon was the result of research with completely another goal. Nevertheless many firms are explicitly or implicitly following a technological trajectory in their R&D (Dosi et al., 1988). In the automobile industry all firms use basically the same kind of combustion engine. So with respect to engine technology they are all in the same technological trajectory. Some manufacturers are currently undertaking research projects to develop a commercially viable electro-engine. It is possible that this will be a new trajectory for the automobile industry in the foreseeable future. In many industries a number of trajectories simultaneously determine the technological core. Diversified firms try to exploit a number of technologies (from trajectories) by making a range of products embodying these technologies for different markets. Sometimes one or a sequence of radical innovations open up a whole area of applications covering several or many industries. Examples of such radical change are microelectronics and biotechnology. These more encompassing 'fields' than trajectories may change the whole way of applying science and technology into commercial use. They are called 'technological paradigms'.

Despite the important role of trajectories and paradigms, diversity among the group of firms in an industry is significant too. Diversity follows from bounded rationality and other behavioral assumptions within evolutionary economics. We briefly discuss routines, firm-specific knowledge and diffusion of knowledge.

Evolutionary economics assumes that decision-makers are boundedly rational and have limited information about market developments and scientific and technological knowledge which can be used to pursue certain technological goals and to select and execute innovation projects. Instead of the optimal choice from a given set of technological opportunities rules of thumb and other routines play a crucial role in acquiring information and making decisions on the basis of limited information and rationality. The moment information and rationality are scarce resources, acquiring information and making choices as rational (in terms of firm goals and internal conflicts) as possible becomes dependent on the characteristics of individuals and the institutional setting of both individuals and firms. As a result firms pursue different technological goals or come up with different results when having the same goals on the base of the same – say public – information. Once diversity occurs path dependency may reinforce such differences. An opposing factor is imitation. Through imitation differences disappear. In many markets 'dynamics' are guaranteed because in the long run many outsiders or small firms come up with radical innovations. In most cases the technological expertise of a firm is a mixture of firm-specific knoweldge and public knowledge. Especially in the domain of innovation firms try to obtain a competitive advantage by doing something different. Consequently R&D is at least partly firm-specific. Within a technological trajectory firms create different 'small' innovations. Firm-specificity becomes more apparent when radical innovations occur.

One of the consequences of diversity, limited information and bounded rationality is that imitation may become difficult. This has significant consequences for the interpretation of the economic value of patents, as we will see in the next chapter. Figures will be shown there about the difference between the costs and time of innovation and imitation.

In Chapter 5 and in the second part the book several theoretical explanations of innovation are given which can be called 'evolutionary'.

The appropriability problem

Technological uncertainty of a project only exists ex ante. If all technological problems are solved then the technological uncertainty of that project is reduced to zero.[13] If we indeed assume that all technological problems have been solved at the time the innovation is applied, then commercial success of the innovator is still far from certain. Market uncertainty exists ex post; by commercially applying an innovation the market will decide whether it will be a success or not. The innovation may be imitated, thus eroding the profits for the original innovator. Another possibility is that one or several other competitors innovate at roughly the same time and do better. A firm which wants to invest in one or a number of innovation projects has expectations about the market uncertainty, and thus about the profits from the innovation. The fact that in most cases getting profits from an innovation project which is yet to start is uncertain, causes the appropriability problem.

From a theoretical point of view a firm as a rational decisionmaker will invest in an innovation project when the expected benefits are greater than the expected costs, and when these net benefits (return on investment, or similar measures) are above a certain level acceptable to the firm. In the previous chapter we saw that such decisions are usually not a clean 'go-or-no go' and regularly one has to decide despite a lack of or highly uncertain probabilities of costs and benefits. Nevertheless it seems plausible that the management team what is able to establish procedures which 'optimize' the decision process and the continuous acquisition of new information about costs and benefits of ongoing projects will in the long run be a more efficient innovator than less rational firms (not including the chance that a firm continuously chooses the right project by luck, or that a given project gives a completely unexpected, but highly commercially successful outcome).

A high degree of market uncertainty may have two quite opposite effects. It may make the firms in the market very cautious, reflected in a strong reduction of R&D expenditure, or it may stimulate them to speed up innovation projects and to look for complex technologies which, for example, may provide them with a niche market. Increased speed and complexity cause R&D expenditure to rise, as the tradeoff curves showed. We already suggested that complexity may result in a niche market position for the innovating firm with a higher chance for monopoly profits. The same holds for speed: the higher the speed of innovation the higher the chance that the firm creates a temporary monopoly position. Low costs and low complexity of innovation may therefore result in low innovation profits, while high costs and complexity may give a higher chance to monopoly profits.[14]

Teece (1986) has developed a concept which gives more insight into the kind of environmental factors which determine the degree to which an innovator can appropriate the profits stemming from its innovation. This concept combines some environmental factors which are exogenous to the innovator with some strategic factors which enable the innovator to enlarge its innovation profits by taking the right actions. Whether or not innovation profits are completely[15] appropriated by the innovating firm depends on three conditions:

- the appropriability regime;
- the life cycle of technology;
- control over complementary assets.

The appropriability regime determines the relative ease with which other firms can imitate the new product marketed (or new process applied) by the innovator. In a 'tight appropriability regime' the innovator is well protected against imitation, in a weak regime not. According to Teece innovators in most industries are confronted with a weak regime and therefore the danger exists that innovators will lose a considerable amount of money to imitators. The appropriability regime depends on two subfactors. First, the legal protection of an innovation, based on patent rights and to a lesser extent brand protection and other legal constraints to imitation. Second, the nature of technology; some technology is easier to imitate than other technologies. Generally new products are easier to imitate than new processes. In Chapter 6 it is shown that in most industries patents do not protect the innovator very strongly, although the patent protection of new processes is better than of new products. The nature of technology is also reflected in the knowledge needed to realize the innovation. Most innovations are a mixture of tacit and codified knowledge. Codified knowledge is that part of the knowledge underlying an innovation which can be written down or stored in another way. An example is the chemical formula of a new product. As soon as a competitor acquires the formula it can imitate this product. Tacit knowledge is that part of the innovation of which the participants of the R&D process are not conscious, but which co-determines the (technological) success of the innovation. An example is the way in which the process with which a new medicine is made in the laboratory is 'scaled-up' to the size required for commercial production. The best way to proceed is to a large extent depending on the experience of the scientists and engineers with former scaling-up projects.

To the extent that market developments permit innovators to choose for process innovations or new products which largely depend on tacit knowledge, a strategy can consciously be followed which 'minimizes' the danger of imitation. Another condition which often gives rise to losing innovators is the life cycle of technology. In a simplified picture the development of an industry can be depicted as the development of a single technology or a set of related technologies over time. Based on the work of Abernathy and Utterback, Teece (1986) distinguishes two stages of technology development in an industry. At the first stage[16] competitors in the market compete with each other in terms of different product designs for the same consumer need (for example the kind of engine in an automobile). At this stage yet another innovator may come on the market with its own new product design. It is clear that as long as the market – that is a considerable number of actual consumers – has not decided about a preferred design, each innovator may be confronted with a loss of innovation profits, especially when a competitor does introduce a design which is soon to be preferred by the market. In the latter case a dominant design occurs. The loss of innovation profits predicted by Teece, when still no dominant design exists in the market, is actually a loss of potential profits when all or part of the consumers would prefer the design of the product of the innovator.

Once a dominant design exists in the market all competitors need to incorporate the technological expertise required for this design; the market shifts from a strong heterogeneity of product designs to homogeneity in terms of a single dominant design.[17] At the level of the industry innovation shifts from product innovations towards process innovations, as Figure 4.5 shows.

The occurrence of a relatively large number of process innovations reflects the need to compete with price and/or quality of the products in the industry, based on the same dominant design. Teece only analyzes industries where this dominant design can be produced on a mass scale. This implies that in the second stage of developent[18] process innovations initially aim at large scale production techniques. In this period of the technology life cycle product innovations merely refine the dominant design or add something to it. At this stage the third condition, control over complementary assets, becomes crucial.

Complementary assets are defined as assets which contribute to the commercial success of the innovation. They can be divided into 'complementary

*Figure 4.5 Product and process innovations during the life cycle of
technology in an industry*

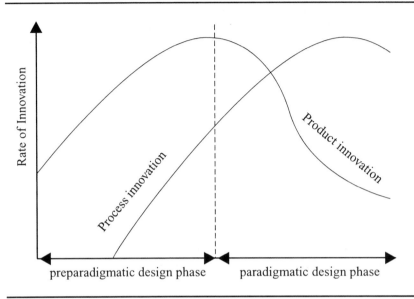

Source: Teece (1986)

technology' and complementary assets for production and distribution of the
new product. Complementary technologies are parts needed to sell the dominant design. An example is a monitor for a new PC based on a new microprocessor. A transaction-cost economics argument comes in here. When complementary technology is standard, for example when the monitor can be used
for several other types of PC, the innovator can appropriate the innovation
profits; when the supplier tries to get part of the profit through re-negotiation
of the contract terms, the innovator can easily contract another supplier. In
contrast, when the complementary technology is specific to the innovation (a
new monitor designed to fit only the new microprocessor), the innovator
becomes dependent on the single supplier who has invested to develop and
produce the new monitor. The supplier may be in the position to bargain for a
nice part of the innovation profits. Here the market power of the innovator and
the supplier of complementary technology becomes important. The example
of several PC producers and their dependence on Intel's microprocessors or
MicroSoft's steering systems is well known. In the same way the innovator
might be confronted with competitors with superior production and distribution facilities. When the product can easily be imitated, these production and
distribution facilities might compensate for the arrears arising from new

product development (again depending on whether these assets are specific to the innovation or not).

In the analysis of Teece's concept it became clear that the benefits of a specific innovation project depend on a number of factors in the firm's environment, part of which can be influenced by the firm. It has not been discussed by Teece whether the expected net benefit of such innovation projects is sufficiently positive to expect a 'go decision'. The point is that perhaps the innovating firm did expect a considerable profit, but that a disregard of such factors as the life cycle of technology and control of specific complementary assets made profit expectations turn out to be overly optimistic. Expected benefits may be positive if the right environmental conditions exist (are to be expected).

The discussion of technological uncertainty and market uncertainty has made clear how difficult it may be for a firm to decide how much money to spend on a certain innovation project. The statement of Link (1993, 1) does therefore not come as a surprise:

> It is important for an organization to evaluate systematically the economic returns that it realizes on all of its investments. R&D is no exception. However, one is hard pressed to find within the literature any single R&D evaluation method used consistently across organizations. One is even harder pressed to find R&D vice presidents (or their counterparts) who are fully pleased with the evaluation method(s) that they use. This may seem odd, owing to the importance of R&D. However, when one thinks about the vagaries associated with research per se, as well as about the collection of individuals necessary to change 'creativity' into value, the difficulty in agreeing upon the accuracy of one single evaluation technique becomes obvious.

Link (1993, 4) uses a specific model of technological development, in which the factors of production 'capital', 'labor', and 'technology' are combined to produce marketable products which create value added. The production factor 'technology' is depending on 'proprietary technology' resulting from R&D of private firms, but also on 'generic technology', 'infratechnology' and the 'science base'. The science base and generic technology are similar to the concepts discussed in Chapter 1. 'Infratechnology research is the process of creating basic scientific and engineering data, measurement methods, test methods, and measurement concepts which increase the productivity of all forms of research, as well as development.' (ibid., 4). It therefore influences the science base, generic technology, and proprietary technology as well as the use of new technology in the economic production process. R&D projects may differ in the extent to which infratechnology, generic technology, and science influence the private R&D of the firm to develop new products and processes as external factors. The precise mixture of internal R&D (technology projects and product development projects) and external technology factors no doubt influences the R&D evaluation method suitable for a specific R&D project.

Economists often use ratio models. Link (ibid., 5) states that:

> By far, the most common methods for evaluating the return on R&D are based on the formulation of some type of input-to-output ratio. Whenever one approaches an evaluation using this type of model, several basic issues arise. First and foremost is the issue of where within the process of technological development the ratio is calculated. Once that is determined, a second issue arises regarding how to measure the inputs and the outputs . . . any ratio model, no matter how carefully the inputs and resulting output are quantified, have fundamental weaknesses. First, it should be remembered that no relationship in the process illustrated in Figure 1 [our Figure 2.1] is linear. In other words, R&D does not lead directly to T. Rather, R&D is one input, along with generic technology and infratechnology, used to produce a stock of technology. In fact, by the nature of infratechnologies the efficiency of the R&D $\rightarrow$ T transformation is dependent upon other technology elements. In a similar manner, T does not directly produce value added. That process depends critically on a product being commercialized and sold. Thus, production and marketing have key roles that must not only be understood but also quantitatively accounted for when attempting to interpret a Value Added/T number (. . .). Second, time is a variable that is fundamental to any evaluation process, but time is absent from Figure 1. R&D in period t will have effects in future periods, and the time lag is likely independent of lags associated with previous R&D efforts (. . .).

An entirely different matter is projects without a (sufficiently high) net benefit. A potential innovation project which is not carried out because it does not have an acceptable net benefit for a private company, may have certain benefits for society. In those cases the benefit of an innovation project for society is greater than for the private firm. An example is a project which is expected to result in a new machine which reduces the pollution of the production process in an industry, but which has higher R&D costs than the costs saved by the one firm not having to pay an environmental agency which induces a tax on pollution. No single firm will undertake such a project, but for society it would be a beneficial undertaking, because pollution is reduced. Another example is an R&D project which does not yield enough benefits for a firm in a specific industry, but which is expected to have substantial 'spillovers' to firms in other industries. One could think of the various commercial applications space technology has brought forward, but which could not have been made appropriable to a private firm which created the innovation from which new innovations sprang off. Basic research is a good example of an area where projects are carried out in which private benefits are considerably lower than social benefits.

We have seen in Chapter 1 that the major part of basic research in OECD countries is carried out or at least financed by government. Nelson (1959) argued in an early stage of the debate that basic research will in most cases not be carried out by private firms, because private net benefits are too low due to appropriability problems. Here we have a specific type of innovation project with high uncertainty. Innovation projects may differ with respect to technological and market uncertainty, as the next section will show.

Theoretically the difference between private and social benefits of an innovation project can be seen as one of the main problems for innovating firms. These concepts reflect several of the other problems firms encounter in practice. We discussed the measurement of private benefits in chapter two. Here we discuss the measurement of social benefits of an innovation project.

Two approaches can be identified in the literature, the 'old' approach, based on the diffusion of an innovation and the new approach, which takes spillovers into account. An example of the old approach is given by Mansfield et al. (1977). They state that the calculation of the social benefits depends on the kind of innovation: product innovations used by (other) firms, product innovations used by households, and process innovations. They give the example of a product innovation used by other firms.

The (net) private rate of return = profit from new product - R&D of the project - production costs (per unit per year...). (Net) social rate of return = profit from new product of firm i + cost savings in the user industry. For a measurement of the cost savings in the industry we need to know the price elasticity of demand of the product of the industry using the innovation (plus old and new prices and profit margin of innovator); 'Rough estimates of n [price elasticity] were obtained from published studies and from the firms.' (ibid., 226)

According to Mansfield (ibid., 235/36) the main factors which give rise to $B_s > B_p$ are:

- market structure: the more competition the less the innovator can appropriate all the social benefits;
- patent protection and the cost of imitation;
- product or process innovation;

In this approach benefits are restricted to direct profits from the innovation and the direct cost savings by using the innovation. Learning is not recognized. The 'new' approach does take such spillovers into account.

Uncertainty and different types of innovation

Freeman and Soete (1997) distinguish various types of innovation with different degrees of uncertainty. Table 4.7 gives a classification of types of innovation according to their degree of uncertainty. This uncertainty is a mixture of market uncertainty and technological uncertainty. Product innovations have a higher degree of market uncertainty than process innovations, because imitation is easier. The more radical the intended innovation differs from what the firms in the market have been doing till then, the higher will be the technological uncertainty. Here a complication resides. What for one firm may be a high degree of technological uncertainty may be a lesser degree for the other. One has to to speak of the degree of uncertainty of an innovation project for the firms on average.

We have seen that a high degree of complexity of an innovation gives a higher chance for monopoly profits, because only a few firms in the market may be able to carry out such a project, and if one of these firms is successful many of the others will not, or will with great difficulty, be able to imitate the innovation (see Table 5.2). Here it comes about that market and technological uncertainty are negatively correlated, at least to the extent that complexity is positively correlated with technological uncertainty. The logic is as follows: the more complex the innovation, the higher the technological uncertainty, the more difficult imitation (the lower the chance that other firms are carrying out a similar project), the lower is market uncertainty for the innovating firm.

In Chapter 3 it became clear that in the management of innovation literature a distinction is made between highly uncertain new technology projects and far less uncertain development projects in which only proven new technology is planned to be incorporated. This implies that technological uncertainty is strongly reduced. It may therefore be concluded that the classification of Freeman differs fundamentally from the types of innovation discussed by Wheelwright and Clarke (1992). While Freeman takes into account many different sources of technological uncertainty, Wheelwright and Clarke exclude the uncertainty of inventions, that is, prototype development. In addition market uncertainty is used differently. Freeman recognizes two main sources of market uncertainty: appropriability problems and the acceptance of the new product by the market. He (as an economist) stresses the danger of imitation and other factors determining the appropriability of innovation profits. Wheelwright and Clarke do not explicitly discuss market uncertainty, but their division of types of innovation, such as new platform products and incremental innovations, derives from 'what the market needs'. Here the emphasis is not on the danger of imitation and the like, but on a strategy to develop the kind of innovations which maximize market share in the long run. Wheelwright and Clarke's classification as such is based on the degree of technological change needed in product and process, which implies a ranking based on techno-

Table 4.6 Types of innovation according to their uncertainty

1	True uncertainty	Fundamental research
		Fundamental invention
2	Very high degree of uncertainty	Radical product innovations
		Radical process innovations outside firm
3	High degree of uncertainty	Major product innovations
		Radical process innovations in own establishment or system
4	Moderate uncertainty	New 'generations' of established products
5	Little uncertainty	Licensed innovation
		Imitation of product innovations
		Modification of products and processes
		Early adoption of established process
6	Very little uncertainty	New 'model'
		Product differentiation
		Agency for established product innovation
		Late adoption of established process innovation and franchised operations in own establishment
		Minor technical improvements

Source: Freeman (1982)

logical uncertainty alone.

The market uncertainty of a specific firm in a market is constituted by the uncertainty of fit between the firm's new product[19] and market demand (the preferences of consumers) and by the difficulty of appropriating innovation profits once the new product is indeed fulfilling consumers' needs (the danger of imitation is an important factor here; this and related issues are further discussed in Chapter 6 on patents). We have seen that high uncertainty innovations may reduce market uncertainty, that is: if a firm innovates, despite a high degree of uncertainty, the chance that innovation monopoly profits will quickly diminish in the market is small relative to the situation of low uncertainty innovations. We will see in the next chapter that the speed of innovation may reduce a firm's market uncertainty, because it gives a lead compared to its competitors. The speed of innovation is clearly related to the organization and management of innovation; topics discussed in Chapter 3. In the subsequent chapter on competition we will present a study on the relation between a firm's organizational structure and specialization and its ability to solve innovation problems (which is more concerned with the nature of technological uncertainty than with its degree), which are caused by the introduction of a new product by a competitor.

As said before, in many innovation models the problem of technological uncertainty is 'solved' by increasing the R&D expenditure per time period. The basic idea is that a unit of R&D expenditure results in the creation of new technological knowledge with which technological uncertainty is reduced and ultimately ruled out. There is, however, a limit to the creation of knowledge per time period, as was implied by the time–cost trade-off function. Therefore the time to reach the required level as given in the 'knowledge producton function' (Figure 4.2) is bound to a minimum.

4 CONCLUSIONS

From an economic perspective the main problem of innovation projects is that the outcome is uncertain. If 'ideas' are taken as included in the number of projects started, only two percent of these projects are a commercial success. Some systematic studies on success and failure of innovations, such as SAPPHO, show that the main sources of failure are managerial and organizational factors. The economics of innovation has largely concentrated on market uncertainty and technological uncertainty as an explanation of failure or a lack of innovation profits. At the industry level differences in technological opportunities may result in differences among R&D intensities within sectors. Failure rates do not, however, vary widely among sectors. Therefore in one industry firms may undertake more projects than in another, but success is largely determined by internal factors.

NOTES

1. Both the mixture of theoretical and practical knowledge and the trial and error content of R&D are dependent on the kind of innovation aimed at. Incremental innovations may be realized 'without' theoretical knowledge and with very few trials and errors.
2. Booz, Allen and Hamilton, *Management of New Products*, 1968 (as referred in Ramsey, 1986), 9–10.
3. Technological uncertainty (tu) can be extended to 'technology projects' where it is defined as the uncertainty whether the project results in 'proven technology' within the planned time and budget. Normally tu will be higher for technology projects than product development projects, sometimes reflected in the absence of a timetable or a sharply defined budget.
4. Profit uncertainty was defined as the uncertainty stemming from the innovation/imitation actions of competitors, while market uncertainty concerned the uncertainty whether the innovation would be accepted by the customer's market.
5. One must realize, however, that many problems, even in innovation projects, are solved in a routine-like way. Usher's combination of existing knowledge and the 'spark of genius' is probably more appropriate for a minority of the problems solved during the innovation process.
6. Note that we do not imply that higher costs are due to more working hours. Two projects may have costed 240 man-hours of work, but one is carried out by two people during three weeks

and the other by four in one week. In the second case there is 50 percent overtime. In practice researchers might have a quite flexible working time in the sense that overtime is not paid at all. To the extent that project execution can indeed be accelerated without an increase of labor cost, the increase in total project cost must be explained in terms of materials and equipment (mis)used (see Mansfield et al., 1972).

7. The analysis is complicated by the fact that the time sequence of approaches influences three variables: (1) The number of periods T required to execute all N approaches; (2) the expected value E(T) of the time required to achieve a successful solution or terminate the project; (3) the expected cost E(C) of the project. These three are functionally related, defining two different time-cost trade-off functions: the relationship between T and E(C), and the relationship between E(T) and E(C).
8. Henderson and Cockburn (1996)
9. Defined as before in terms of the probability that the task can be completed within the budgetted time and cost.
10. Mansfield et al. (1972, 145) distinguish the following five stages: applied research, specifications, prototype or pilot plant, tooling and manufacturing facilities, manufacturing start-up.
11. If we define the time period over which a stage is completed as t_i and t_{i+1} respectively, the overlap x used by Mansfield must fulfill the following condition: $0.3t_{i+1} < x < 0.7t_{i+1}$.
12. See Mansfield (ibid., 152). The number of innovations studied is, however, limited and the relationship with cost could only be examined for one stage.
13. Practice could be a bit more complicated. Some innovations turn out to have unexpected technological problems after they have been commercially applied.
14. This does not neccesarily mean that the highest speed and complexity of one's project compared to the whole industry is the best strategy. A 'fast second' strategy may be better, because it avoids some of the technological and market uncertainties the first innovator is confronted with.
15. We have already discussed the possibility than a particular innovation may give rise to a whole chain of subsequent innovations. In a broad sense one could argue that the profits from the chain of innovations also belong to the original innovator, at least partly. Teece (1986) discusses the limited problem of appropriating the profits from a single new product in the market.
16. Which is called the 'prepradigmatic stage'.
17. Of course, an industry at a certain moment, may be typified by a number of dominant designs, if the main product is complex and consisting of a number of subsystems.
18. This stage is called the 'paradigmatic stage'.
19. The fit of the firm's process innovation and market demand must necessarily run via its effects on the product(s): product quality improvement, cost price reduction, decrease of repair, and so on. due to the change in the production process.

REFERENCES

Booz, Allen and Hamilton (1968), Management of new products, New York

Dosi, G. et al. (1988), *Technical Change and Economic Theory*, London, Pinter

Freeman, C. (1982), *The Economics of Industrial Innovation*, London, Pinter

Freeman, Christopher and Luc Soete (1997), *The Economics of Industrial Innovation*, London, Pinter

Henderson, Rebecca and Iain Cockburn (1996), Scale, scope, and spillovers: the determinants of research productivity in drug discovery, *Rand Journal of Economics* 27, 1, 32–59

Jaffe, Adam B. (1989), Characterizing the technological position of firms, with application to quantifying technological opportunity and research spillovers, *Research Policy* 18, 87–97

Kamien, M.E. and N.L. Schwartz (1982), *Market Structure and Innovation*, Cambridge

(MA), Cambridge University Press

Link, A.N. (1993), Methods for evaluating the return on R&D investments, chapter 1 in B. Bozeman and J. Melkers (eds), *Evaluating R&D Impacts: Methods and Practice*, Dordrecht, Kluwer Academic Publishers

Mansfield, E. (1968), *The Economics of Technological Change*, London, Longmans

Mansfield, E., J. Schnee, S. Wagner and M. Hamburger (1972), *Research and Innovation in the Modern Corporation*, London and Basingstoke, Macmillan

Mansfield, E., J. Rapoport, A. Romeo, S. Wagner, and G. Beardsley (1977), Social and private rates of return from industrial innovations, *Quarterly Journal of Economics*, 221–39

Nelson, Richard R. (1959), The simple economics of basic scientific research, *Journal of Political Economy* 67, 3, 297–306

Nelson, R.R. and S.D. Winter (1982), *An Evolutionary Theory of Economic Change*, Cambridge (MA), Belknap Press

Ramsey, Jackson Eugene (1986), *Research and Development*, Ann Arbor, Michigan, UMI Research Press, University Microfilms

Scherer, F.M. (1984), *Innovation and Growth; Schumpeterian Perspectives*, Cambridge (MA), MIT Press

Teece, D.J. (1986), Profiting from technological innovation, *Research Policy* 15, 286–305

Twiss, Brian (1992), *Managing Technological Innovation,* London, Pitman fourth edition

Wheelwright, Steven C. and Kim B. Clarke (1992), *Revolutionizing Product Development*, New York, Free Press

5. The nature of technological competition

1 INTRODUCTION

In the previous chapter an explanation has been given of the behavior of innovators and the uncertainty with which it is surrounded. It turned out that one important factor was the market in which the innovator operates (or which he wants to enter). The main elements to characterize a market in an economic sense are the size and nature of demand, and the nature and degree of competition. We concentrate on the nature and degree of competition.[1] A core element of innovation and competition is the appropriability of innovation profits. Getting the benefits of an innovation or not is one of the the main problems of an innovating firm, and usually the competitors cause the appropriability problem. The micro factors discussed in the previous chapter are strongly linked to it. Here we concentrate on interactions between firms in a market and the overall result at the market level, instead of the influence of the market on an individual firm's decisions and performance.

In this chapter the empirical investigation concentrates on the way firms react to the technological performance and/or behavior of competitors. The theoretical section relates to a very broad body of literature. The main source for theoretical elements discussed here is 'industrial organization' and to a lesser extent 'evolutionary economics'. In the industrial organization literature the topic on innovation is headed 'market structure and innovation'. The appropriability problem is also a core element there. In Evolutionary Economics the interaction between firms in a market is equally important. In this theoretical approach the selection of innovations is as much a core element as appropriability is.

2 THE EMPIRICAL SIDE: THE IMPACT OF TECHNO-LOGICAL COMPETITION ON INNOVATIVE BEHAVIOR OF FIRMS

To understand better the results of empirical studies about competition and innovation, a distinction must be made between an innovating firm reacting directly to R&D investment decisions or innovation results of another firm in the market, and a firm's reaction in terms of R&D and innovation (strategy) to an increase in competition in its market(s) more in general.

Looking at the vast amount of theoretical studies on technological competition it is surprising to find that we know relatively little about how competitors in various markets react directly to each other in terms of innovation strategy and R&D expenditure. Much more is known about the impact of (a specific degree of) competition on the average level of R&D investments and the innovation results in the market. One can speculate about the insignificance of the 'direct approach'. Perhaps economists 'automatically' assumed that, while technology is a core element of competition, firms will react sharply on a competitor's technological succes by increasing their own R&D. It turns out that they do not, or at least, sometimes they do sometimes they do not. This result must be taken into consideration when discussing the theoretical contributions on this topic.

One of the older systematic accounts of the topic is about the reactions of the US steel firms on the introduction of the 'basic oxygen furnace' (BOF) which was introduced in 1952 by VOEST in Austria. A small American firm adopted the BOF in 1954, but despite a history of evidence of the value of this new process, the two largest US steel producers did not have BOF in process until 1964. Meanwhile US crude steel production as a percentage of world steel production fell from 39 percent in 1955 to 25 percent in 1962 (Gruber and Marquis, 1969, 43–4). Figure 5.1 shows how the US steel industry reacted on the loss of market share by gradually increasing its R&D intensity.

Figure 5.1 The United States steel industry: challenge and response

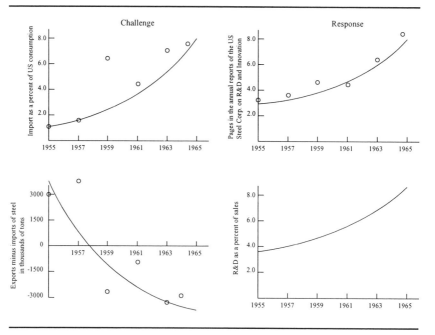

Source: Gruber and Marquis (1969)

From the example above it became clear that American steel firms knew about the introduction of the BOF in Europe. How quickly do firms usually know about innovation plans and implementation of new technology by competitors? According to Mansfield (1985, 217) we know little about this subject:

> To understand the process of industrial innovation and how rapidly innovations are imitated, it is obvious that economists should study the nature and extent of the information that firms have about their rivals' technology and R&D programs. Yet there have been no systematic empirical studies of the speed at which various kinds of technological information leak out to rival firms.

In his study about 100 US (large) firms Mansfield (ibid., 218–20) found that development plans are on average in the hands of at least some of the rivals within about 12 to 18 month after the decision has been made, and competing firms know about the detailed nature and operation of the new product or process developed by the firm within about a year.

The next question is how do competitors react on such information? One of the most extensive studies dealing directly with how firms react on each other in terms of R&D investment is offered by Scherer (1992). The background of this study is the loss of the United States' dominant position in terms of science and technology. This weakened position of US firms is reflected in both the technological dominance by non-US firms in several industries and the decline of technological performances by US firms. Scherer (1992, 1) states that:

> Other nations . . . industries moved visibly ahead [of the US] in automobile design and quality, steel-making technology, consumer electronic goods, shipbuilding, textile machinery, general-purpose numerically controlled machine tools, copying machines, ceramics, high-voltage electrical transmission and rectifying apparatus, nuclear power reactors, industrial enzymes, and much else.

With regard to the technological performance of US firms Scherer stresses that company-financed R&D grew at a lower pace from 1970–90 than in the 1960s, while basic research financed and conducted by private firms even diminished after 1966 and 'failed to surpass its 1966 level again until 1981' (ibid., 3). Moreover, the output of this reduced R&D growth diminished in the same period:

> The number of US invention patents issued to US corporations (. . .), which had been growing at an average rate of 4.3 percent during the 1950s and 1960s, peaked at 43,022 in 1971 and then declined to less than 30,000 per year on average in the 1980s. (ibid., 3)

According to Scherer this had notably negative effects on the productivity growth and export performance of US industries. The question is how then did US firms react on this technological challenge of foreign competitors? The empirical part of the study consists of eleven case studies on specific products and statistical analysis of data on 308 firms. The cases concerned wet shaving apparatus, television and VCR sets, automobile and truck tires, electronic calculators, commercial airliners, cameras and film, diagnostic imaging equipment, digital switches, optical fibers for cables, facsimile machines, and earth moving equipment. Scherer (ibid., 108) concludes that:

> US manufacturers have reacted in widely differing ways to new technological challenges from abroad. Some, such as Gillette, Eastman Kodak, General Electric (in digital imaging), and Corning-AT&T (in fiber optics), redoubled their own R&D efforts in response to growing high-technology competition. Others, notably the makers of such consumer electronic apparatus as television sets and video recorders and the early market leaders in facsimile machines, in effect surrendered the market to overseas rivals. Still others, such as the calculator and tire manufacturers, AT&T Technologies (in digital switches), and Boeing, continued vigorous R&D efforts but ceded sizable market niches.

In the statistical analysis of the study the impact of foreign competition on the firms' R&D expenditure was estimated. The firms were located in industries where R&D does play a sizable role.[2] The results were the following (ibid., 171–2):

> As in the case studies, we find considerable idiosyncrasy in companies' competitive reactions. But import competition does appear to have made a difference. The short-run reaction to increased imports (or declining net exports) was on average submissive; that is, R&D/sales ratios fell. However, large, diversified firms occupying concentrated markets reacted more aggressively than their smaller, less diversified counterparts. Multinationals reacted more heterogeneously to rising imports and less submissively to net export declines. Insulation from import competition through trade barriers blunted firms' reactions. Over the longer run, there appears to have been a reversal of the average reaction pattern from submissive to aggressive, although the evidence on this point remains weak.

The main point is that most firms, at least in the short run, reacted on increased technological competition from abroad by not increasing or even cutting back R&D expenditure. Not increasing R&D could be a matter of not being able to come up with more promising innovation projects or of not being able to finance more projects. In the above quote there is some evidence that at least part of the firms may already have had detoriating profits, because of increased foreign competition, which unabled them to react properly. To the extent that diversified firms in concentrated markets have higher profit margins than other firms, they may have been able to finance an increase of R&D expenditure, while smaller firms were not able to react with an increase in R&D. Two minor conclusions of the study are of interest in this respect. First, firms headed by technologically educated individuals tended to react more aggressively to rising imports (ceteris paribus). Second, the reactions were more submissive in industries where basic academic research in physics, chemistry, and biology is important. This kind of basic knowledge tends to be public, so it is more difficult for firms in the domestic industries to appropriate innovation profits, because foreign firms learn quickly about basic discoveries by US universities.[3]

In summary two types of characteristic of an industry seem to influence the firms' reactions on innovations by foreign rivals: the degree of concentration in the domestic market and the kind of knowledge on which innovations are based. As has been discussed by Teece (1986) the kind of knowledge co-determines whether the innovator can appropriate innovations profits. The more industry- or even firm-specific the knowledge underlying innovations, the more firms can rely on 'being the first in the market'. According to Scherer science-based industries, as classified by Pavitt (1984), offer a weak appropriability regime.

The second conclusion on market concentration relates to the dominant line of research in 'market structure and innovation': what is the effect of the size and/or market power of the firm on its innovation performance? We discuss this matter in the next section.

The study by Scherer focuses on the reactions by American firms on import penetration by Japanese firms. What about a firm's reaction to an increase of R&D or innovation results of a domestic rival? Or put more generally, how does one firm react to innovation by another firm in the market? Cockburn and Henderson (1994) give an account of a study in the pharmaceutical industry. The interesting perspective of their empirical research is that on the one hand it is not industry-wide but restricted to antihypertensive drugs' and on the other hand it is relatively detailed, as data are available on both research projects and whole 'antihypertensive' programmes. Their results show that all ten firms in the sample[4] continue to publish in the 'open research literature' after the initial burst of scientific interest in this field. This does not indicate the kind of secrecy one would expect in 'a winner takes it all' strategy. Furthermore it is stated (ibid., 495):

In general, the managers whom we interviewed claimed that in planning their investment programmes they focused on three criteria: the size of the unmet medical need, the scientific potential of a field, and the idosyncratic capabilities of their researchers . . . when pressed on the role of competitive investment in their criteria, the managers that we interviewed claimed that they tried to avoid head-to-head competition, or 'racing', as inherently unproductive.

This means that a firm is not investing in a specific project because a rival has already done so, nor does it stop a specific line of research once a patent has been granted to another firm. In fact, the importance of spillovers inside the firms of knowledge from one programme in the field to another and of spillovers between firms is one of the main findings of this study; 'rivals' R&D results are positively correlated with own research productivity, which we interpret as evidence for extensive R&D spillovers rather than the depletion externality implied by winner take all models'.

Meron and Caves (1991) found that in a sample of 28 US manufacturing industries, leaders and followers reacted positively to each others' increases in R&D expenditures, while fringe firms' investment decreased with their large rivals' investment. This last conclusion seems to be consistent with Scherer (1992). Indirect evidence is provided by Hundley et al. (1996), who investigated the presumed difference between Japanese and US firms with respect to the relation between profitability and R&D intensity. In Japan a positive correlation exists between firm profitability and R&D intensity, in the US there is a negative correlation. For research-intensive industries in the US, however, a positive correlation is found. Birnbaum-More et al. (1994) argue that product and process innovation only account for 20 percent of the total actions taken by firms in reaction to changing market shares compared to important rivals. When concentration in the industry increases and the annual growth rate of aggregate demand decreases (which usually stands for matur-

Table 5.1 Response time of competitors according to firms which introduce new products

Competitor's response time	N	Percentage of firms
Less than 6 months	49	13.4
6 month – 1 year	66	18.0
≥ 1 year	105	28.7
No visible response	146	39.8

Source: Bowman and Gatignon (1995, 46)

ity of the market) process technology becomes more important. In contrast to the technology life-cycle theory, they found that product technology as a strategic action grew in significance with increased concentration and decreased compound annual growth rate.[5] Bowman and Gatignon (1995) investigated the determinants of competitor response times to a new product introduction, based on data of 101 strategic business units from the PIMS database. Although industry concentration did not come out as significant, it is interesting that industry growth was positively correlated with response time. The response time itself varied from 'one year or longer' to less than 6 months. sixty percent of the competitors reacted to the firms' introduction of a new product.

One way to respond is by trying to imitate the newly introduced product (see Mansfield).

Although scarce, the empirical evidence indicates that firms can both increase or decrease their R&D investment when a rival increases its investment, but the likelihood that a positive relation between competitors' R&D investment exists is augmented when firms are large and have significant market shares. The same holds for research-intensive industries, like pharmaceuticals.

The effects of concentration and size

In the presentation of empirical results about firms' reactions to a competitor's increase in R&D it was found that large firms in concentrated markets are more likely to increase their R&D than smaller firms. Most of the empirical research has been devoted to the impact of concentration in the industry and size of the firm(s) on R&D investment and innovativeness (R&D productivity). An overview of this market structure and innovation literature can be found in Baldwin and Scott (1987) and Kamien and Schwartz (1981). Scherer and Ross (1990) and others have concluded that the empirical evidence is rather inconclusive. Although monopoly and pure competition as extreme types of market structure, related to concentration and average firm size, are clearly unfavourable in most situations, which in-between degree of competition gives the best results in terms of R&D intensity and innovation output of the firms in the industry is unclear. Pavitt (1982) discusses this matter in a way which is quite interesting for the purpose of this book, because he relates R&D productivity and firm size with the nature of the innovation process in small and large firms. We deal with this study in the theoretical section.

Originally the relationship between market concentration and innovation (both inputs and outputs) was measured as an inverse U-form at the level of lines of business or more rough classifications of industries. According to Levin et al. (1985) 'The effect of concentration, however, was sharply attenuated when we controlled for interindustry differences in technological opportunity and in the appropriability of returns from new technology'. The same

Figure 5.2 The effect of firm size and market concentration on R&D intensity

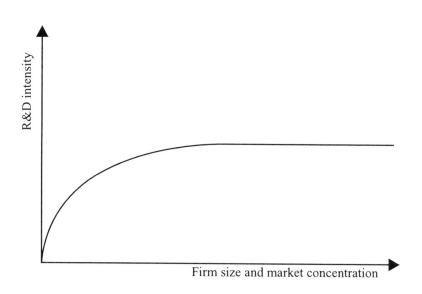

holds for the positive effect of firm size and the size of business unit on R&D intensity (Cohen et al., 1987). Today the effect of market concentration and firm size on R&D intensity is considered to be minor; that is, beyond a certain threshold there is no sizeable positive effect, as Figure 5.2 shows.

3 THEORETICAL ASPECTS OF TECHNOLOGICAL COMPETITION

According to Scherer (1992) economic theories offer little support for the empirical evidence that, on the one hand, most firms do not increase R&D when confronted with increased technology competition and that, on the other hand, reactions are quite diverse. Industrial Organization and the 'New Trade Theory', more specifically the game theoretic approaches within the two fields, are able to predict every R&D reaction (depending on the specific behavioral assumptions and the pay-off structure), but they have not systematically confronted theoretical predictions in a specific setting with empirical outcomes.

The nature of technology in a market may strongly influence the intensity and nature of competition between its firms, as the empirical section of this chapter showed. How the technology of an industry is perceived and how this relates to the view on the competition process is fundamental. We begin this section by arguing that different theoretical visions may contain fundamentally different ways in which technology and market competition interact. To make this clear we use the earlier mentioned distinction between a market with homogeneous technology and a market with heterogeneous technology. We elaborate on this now.

Let us assume that a firm consists of only a production unit and an R&D unit. In the homogeneous market the production units of all firms make products with similar technological and demand characteristics. Similar demand characteristics mean high cross-elasticity of demand between the products of the firms. Similar technology means that the technology of the products and the production line with which they are produced are similar for all firms. The R&D units all focus on either an improvement of the existing (similar) technology or on a similar innovation project to create a radical change.[6]

In the heterogeneous market the products of the firms have high cross-elasticities of demand, but lower than in the case of the homogeneous market, allowing for some product differentiation. More importantly, the firms may differ with regard to the technology incorporated in their products and their production process. By implication, the technological or knowledge base of the firms may differ. Consequently, R&D units of different firms may undertake innovation projects which differ in the kind of technology aimed for. For simplicity we assume that each firm coincidently spends amount X on R&D, although the projects may differ substantially.

The two market visions have different implications. In the homogeneous case technological competition is 'immediate'. If one firm succeeds in creating and marketing a new product, the market share of the other firms will immediately decrease. If a few or all firms in that market are undertaking innovation projects and if the 'winner' gets a watertight patent[7] they can be said to be in the same race and real monopoly profits are obtained by the innovator. Consequently, the market share and profit of the other firms will diminish. On the other hand, when patent protection is weak, imitation is relatively easy because the technology of the firms is similar. In that case market uncertainty for the innovating firm is very high and it may hesitate to invest in R&D, given a sufficient farsightedness to anticipate on appropriability problems with respect to its innovation profits.

Private benefits of an innovation project of firm i in the homogeneous market are thus strongly dependent on the results of the other firms' R&D. In the case of one winner (strong patent protection) the R&D expenditure of the other firms has been a waste.[8] In theoretical terms: the private costs of innovation are the amount X of the single innovator who gets the patent; the social costs of that firm's innovation project may equal nX, where n is the number of firms in the market and X the R&D expenditure of each firm.[9] In a 'winner take all' situation firms are stimulated to increase R&D from X to X + y (as we have seen, speeding up an innovation project increases R&D for the individual firm). In contrast, when there is not a single winner, R&D investment will on average be less for each firm, say X - y.

In the theoretical literature the danger of overinvestment in R&D in a 'winner takes all' situation has been extensively discussed. Included in the discussion is the possibility that a weak patent protection and farsightedness of actors may result in a 'stalemate', where no firm, or too few firms, invest in R&D (see Reinganum, 1989 for an overview).

In the heterogeneous market the appropriability problem is less severe, because some product differentiation is possible from the start and each firm's innovation project may have some future niche market, even when patents give a strong protection. As a result it can be expected that the R&D expenditure of each firm remains X, independent of the innovation performance of the various firms in the market. This implies that for this market holds: Cs = nX. Consequently the social costs of the heterogeneous market may be higher or lower than for the homogeneous market (where social costs were n(X - y) or n(X + y). In addition the social benefits in both markets will probably differ. In the homogeneous market the benefit is equal to the increased welfare due to the new product for the consumers.[10] In the heterogeneous case a number of new products may enter the market, hence technological diversity is higher than in the first case. This gives society more choice and will increase the probability that one of the new products turns out to be 'better' than in the homogeneous case.[11]

Market structure and innovation

The topic of market structure and innovation belongs to an approach in 'industrial organization' called 'structure–conduct–performance'. One can distinguish a theoretical and an empirical branch. The theoretical branch is closely related to the microeconomic theory of the market, and deals largely with firms' behavior and action as if they can be represented by a single rational decision maker who is well informed about different possible alternatives for investment. The empirical branch started with the work of Bain (1956) in which, although implicitly, bounded rationality and limited information of firms play some role.

The core issue of the structure–conduct–performance approach to innovation is the question as to whether a specific degree of competition results in the best innovation performance of the firms.[12] This degree of competition is incorporated in the term 'market structure'.

Traditionally this approach has been occupied with the question as to whether the market structure of monopoly or that of strong competition gives the best innovation performance. As always the answer is dependent on specific assumptions, in this case about the effect of monopoly and strong competition on R&D expenditure and innovation performance. Under most conditions strong competition results in better performances than monopoly. An exception is the case where innovation depends on considerable economies of scale in R&D. A weak point of these analyses is that 'strong competition' is an ill-defined concept. How strong is 'strong competition' precisely?

Empirical research in the field (see the empirical section of this chapter) as well as more refined theoretical analyses (for example, Reinganum, 1989) concluded that market structures between monopoly and strong competition give the best innovation results for the firms on average. This means that market structures like tight or loose oligopoly, and monopolistic competition are the most favorable. The following points enable a better judgment to be made of the practical value of this important theoretical insight: (1). The majority of current markets in the industrialized world can be typified by the abovementioned market structures with an 'intermediate degree of competition'. This implies that we need a more refined analyis in order to find out which intermediate degree of competition gives the best result. How tight or loose, for example, must an oligopoly be? (2). There is usually no explicit choice between the homogeneous market model or the heterogeneous model as the theoretical underpinning of the above conclusions. As we have seen in the introduction to this theoretical section, these models may have different implications with respect to R&D costs and innovation performance (innovation benefits).

Many of the studies mentioned rely on limited data of innovation activities. They are quite often based on R&D alone or, when they do include output measures, these are usually confined to patents. Patents are generally seen as limited proxies for innovative output (see the next chapter for a further discussion). No, or insufficient, attention is paid to innovation capacity, to the firm's ability to realize innovations with a certain amount of R&D. Mowery and Rosenburg (1989, 5–6) have stated as follows:

> Neoclassical theory's emphasis on the appropriability of the results of research must be complemented with an analysis of the conditions supporting the utilization of these results . . . Utilization of the results of research is heavily influenced by the structure and organization of the research system within an economy, a topic on which the neoclassical theory is either silent or incorrect.

The measurement of R&D productivity has been of concern to the literature on market structure and innovation, but the determinants of this productivity have been neglected (such as the influence of the management and organization of innovation activities, discussed in Chapter 3). Evolutionary economics (see p. 135) tends to include these kinds of topics.

As discussed earlier, Pavitt (1982) examined the issue in relation to characteristics of the innovation process itself. He starts from the 'accepted fact' that empirical studies on R&D expenditure of firms reveal that R&D intensity increases with firm size until a certain threshold point. After this point R&D intensity remains constant (see Figure 5.2 in the empirical section). Patents as a percentage of R&D have been decreasing with firm size. (Soete, 1977) This poses the question of whether the R&D productivity of large firms is lower than that of smaller firms. It also relates to the phenomenon that R&D productivity has been decreasing over the last decades. (Scherer, 1992)

Pavitt (1982) interprets these facts differently. He argues that the number of patents per unit of R&D input and R&D intensity of firms are a function of the nature of their innovative activities. As a result the measurement of both innovation input and output may lead to different results for small and large firms.

With regard to input measurement, Pavitt (ibid.) argues that small firms may undertake more R&D than is officially registered, while large firms are more specialized and formalized in their innovation activities. As a result they register more activities as R&D than small firms. In addition they undertake more diverse innovation activities which are not carried out by small firms, or when they are, they do not register them as R&D. Examples are process improvements in engineering and large test facilities.

With regard to output measurement, some of the activities which large firms register as R&D do not result in patentable outcomes, whereas patents can be granted for the results of most of the R&D in small firms. Examples are basic scientific research and complicated test procedures. Consequently small firms have relatively more patents than large firms. The argument is therefore that innovation output of large firms is underestimated and the innovation input of small firms is underestimated. By implication the R&D productivity of large firms is underestimated and that of small firms overestimated.

In the debate on market structure, size of firms and concentration, several other interesting lines of research can be mentioned. Henderson and Cockburn (1996) argue that many micro-studies are carried out at the level of aggregate R&D investment of firms in the industry. Their study on the pharmaceutical industry shows that the productivity of research projects in large firms is better than in smaller firms, but more detailed analysis revealed that it was not economies of scale that explained this phenomenon, but economies of scope; the broader the R&D fields in terms of different programs and number of diverse projects per program, the higher the chance that intrafirm and interfirm knowledge spillovers increased productivity of single projects.

Another point is 'what is a small firm?' Pavitt (1982) argued that data on small firms may underestimate inputs and overestimate outputs compared with large firms. In the section of market structure and diversity below we discuss another study on small versus large firms which might have a different definition of size of firm. Acs and Audretsch (1987, 1988), Kleinknecht (1989) and others have studied innovation, R&D and firm size extensively, often based on surveys containing more information about firms between 10 and 500 employees than the usual databases. They confirm the earlier mentioned results that small firms are not less R&D productive than large firms. More specifically Acs and Audretsch look for conditions under which large and small firms are the most innovative. They conclude that (Acs and Audretsch, 1987, 573):

> Industries which are capital-intensive, concentrated, and advertising-intensive tend to promote the innovative advantage in large firms. The small-firm innovative advantage, however, tends to occur in industries in the early stages of the life-cycle, where total innovation and the use of skilled labor play a large role . . .

An interesting question is whether small firms (<500) are only successful in industries where relatively few economies of scope and scale in R&D exist. This question has not been addressed frequently. Acs and Audretsch (1987, 273) provide indirect evidence. In addition to industries in the early stages of the life cycle, small firms are also performing well in industries where concentration is high: 'the less an industry is composed of small firms, the greater is the relative innovative advantage of those existing small firms over their larger counterparts'. This could be interpreted as the viability of an innovation strategy, under the pressure of the dominating large firms in the industry (in terms of market share), not based on economies of scale in R&D. In Section 4 we discuss how the innovation strategy of small firms in an industry may differ from that of their large rivals.

Market stimulus and market room effect

The better innovation performance of a strong competition structure over a monopoly structure lies in the stimulus effect of competition. In the extreme case[13] a monopoly does not provide any incentive (stimulus) to the single firm to innovate, because a new product replaces the monopolist's old product (this is called cannibalization). The R&D costs of the new product can be incurred through a higher price of the new product compared to the old, but depending on the price elasticity of the demand of the monopolist's product, the total profit in the market may even go down. In case the total profit of the monopolist increases, the crucial question is whether the monopolist's R&D costs, production costs, and other costs are higher than in the situation where competition exists. Normally, the argumentation of economists is that due to X-ineffiency the average costs per unit of production of the monopolists are higher than in the case of competition[14] and that these higher costs lead to higher prices. From society's point of view this is clearly an unfavorable situation because – in the welfare framework of microeconomics – there are social costs (equal to the monopolists R&D costs) but no social benefits, because there is an increase in the producer's surplus but a decrease in the consumer's surplus (less consumers are buying the technologically more advanced product[15]). In the case of competition the firms are forced to spend more money on R&D, which increases the speed of innovation, and the price of the new product will be lower than in the monopoly situation[16]. The increased level of R&D caused by competition is called the market stimulus effect of innovation. As we have seen, such an effect holds for both the homogeneous and heterogeneous market case. There is, however, a limit to the positive effect of competition on R&D expenditure. Depending on the level of R&D required to compete in innovation in the market, the future sales in that market could be insufficient to let all firms capture enough return to get their investment in R&D (and additional investment in production, distribution, advertising, and so on. due to the new product) back. We give an example to further clarify this.

Assume that a market at a certain moment is characterized by the following. In the current year ten firms exist in the market and there is one product sold (homogeneous case). Each firm has sales per year of 100, profit of 10, and R&D expenditure of 10. For simplicity we assume that only R&D needs to be financed from the profit of the firm and that somehow profit in year t can be used to finance R&D in year t. Assume that each firm needs to develop and market a new generation of the current product every year in order to stay competitive, and that the R&D expenditure rises by 25 percent per year. Total demand in the market grows by 10 percent per year. The amount of R&D expenditure of 10 in year 1 allows every firm to come up with a new generation product in year 2 (we assume that no technological uncertainty exists). In year 2 the average sales per firm are 110 and the required R&D 12.5. If the profit margin stays the same, profits per firm are 11. On average there will already be some problem with financing R&D by internal means. In year 3 sales are on average 121, profits 12.1 and R&D has risen to 15.6. In year 2 the total profit in the market to finance R&D is 110, while total R&D required for the 10 firms is 125. In fact the market has only room for nine firms in year 2 (when total profits remain 110 due to decreased competition, they can be used to finance the required 9 x 12.5 = 112.5 R&D). In year 3 total profits are 121 and the required R&D per firm has risen to 15.6, which results in a total of 124.8. The market room is reduced to less than eight firms, if one wants to avoid losses.

The evolutionary perspective

In contrast to the neoclassical approach of the market,[17] the evolutionary approach has been occupied with technological change from the beginning. Schumpeter is commonly seen as the predecessor of this approach. Most models of the market of evolutionary economics assume that firms are boundedly rational decision makers, and because of that it is important to understand the procedures according to which they make decisions.

Nelson and Winter (1982) can be seen as the first evolutionary economists who came up with a comprehensive theory of the market combined with a theory of the firm which offers an alternative to the neoclassical microeconomics model of the market. In their theory of the firm a distinction is made between routine behavior and deliberate search. This distinction is important for an understanding of innovation and the use of technology in this approach.

The evolutionary theory of economic change, like neoclassical theory, focuses on the market as a whole. In contrast to neoclassical theory the evolutionary theory uses quite different behavioral assumptions, and innovation and R&D are core elements.[18] As in evolutionary theory in biology, selection and mutation (in this case a change of the market product and/or the organization of the firm) play a dominant role. The new products introduced by the firms are evaluated by the early clients and ultimately 'the market' selects the best products. In the economy mutations, that is new products or other innovations, do not occur completely at random but are the result of deliberate action and good fortune together. The discovery of nylon was unintended; the researchers at Dupont were undertaking a project with quite a different purpose. In other cases firms deliberately focus on the replacement of an existing product.

As mentioned before, according to the evolutionary perspective most firms operate in a technological trajectory. Such a trajectory consists of a core technology on which the product or a series of products are based. This core technology or dominant design has been the result of a selection by the market of various designs from different firms. In practice firms in an industry may channel their R&D according to a number of trajectories. An example of a dominant design is the combustion engine in the automobile industry. Through the market introduction of a radical innovation one of the technological trajectories of the industry may change. This would be the case when one of the automobile manufacturer would introduce a commercially viable electro-engine. In the same way the invention of the jet-engine meant a new technological trajectory for the airplane industry. Without thorough knowledge of this new technology firms could not survive in most segments of the industry.

Despite the assumption that firms in an industry operate in the same technological trajectory much attention is paid to diversity. Although the example of competition in the 'machine industry for chips' later in this chapter will show that firms often stick to the technology they are used to and good at, other technological developments almost always occur, first, because new firms enter the industry (we saw the radical innovations often originate from outsiders), and second, because each incumbent firm in the industry differs from the other. This last point relates to bounded rationality and related behavioral assumptions. Because firms have imperfect information about technological developments elsewhere, because their employees have various backgrounds and experience in R&D and because each firm has its unique procedures, problem solving methods, and so on, they come up with different innovations. And because of bounded rationality firms use rules of thumb and other routines to make decisions. In contrast neoclassical theory assumes perfect information about technological alternatives and the abilities to always select the best innovation project. This leaves little room for diversity.

According to the evolutionary perspective in economics every firm has a unique knowledge base. Through imitation and public knowledge there is a tendency of firms to have the same technological knowledge and undertake similar innovation projects. Through surprise and creativity firms try to do something different from the others in the market and gain a competitive advantage. Differences between firms make imitation more difficult than in a situation where firms in an industry are similar in any respect. In the next chapter we will see how those different views influence the analysis of the impact of patents (which are there to make imitation more difficult).

4   DIVERSITY AND MARKET STRUCTURE

Some recent studies try to combine insights from evolutionary economics with those from the empirical branch of industrial organization. We are especially interested in the incorporation of some forms of diversity of firms' strategy, organization, and so on, to understand the technological developments in the market. We discuss two examples here. The first one makes a difference between large and small firms in a market with respect to technological change (Cohen and Klepper, 1992), the second one (Henderson and Clarke, 1990) relates the pattern of technological change in a market to specific organizational problems of the firms.

The different role of small and large firms

Some industries consist of relatively large firms, other of relatively small firms. In many industries a considerable variation exists in the size of the firms. One explanation for this phenomenon has been that small firms are a

Table 5.2 The costs and time of imitation

Type of innovation / Percentage	< 25	26–50	51–75	76–100	> 100	Timely duplication not possible
New Processes						
Major patented new process	1	5	10	66	26	10
Major unpatented new process	5	10	55	49	6	2
Typical patented new process	2	15	61	41	6	2
Typical unpatented new process	8	43	58	14	4	0
New Product						
Major patented new product	1	4	17	63	30	12
Major unpatented new product	5	12	58	40	7	4
Typical patented new product	2	18	64	32	9	2
Typical unpatented new product	9	58	40	15	5	0

Type of innovation / Time in months	< 6	6–12	13–36	37–60	>60	Timely duplication not possible
New Processes						
Major patented new process	0	4	72	37	9	7
Major unpatented new process	2	20	84	17	2	4
Typical patented new process	0	40	73	13	0	3
Typical unpatented new process	8	66	47	6	1	1
New Product						
Major patented new product	2	6	64	40	8	9
Major unpatented new product	3	22	89	12	1	2
Typical patented new product	5	39	72	6	4	3
Typical unpatented new product	18	67	39	4	1	0

Source: Levin et al. (1987)

remainder of the 'old' situation in the industry, when firms were smaller and competition less severe. The explanation offered by Cohen and Klepper (1992) is more positive with respect to small firms.

The main element of Cohen and Klepper's explanation is diversity. Each firm in an industry may have its own approach to innovation. More generally it can be assumed that an industry at a given moment has N different approaches to innovation, which may be partly unknown to the firms. The approaches could consist of different ways to process innovation (large scale or automation, for example) and different product innovations. Each firm chooses M approaches to innovation (M ≤ N). Because of ex ante uncertainty about each approach, firms may come up with different choices. When each approach requires the same expected amount of R&D[19], R&D intensity of a firm in the industry depends on M. M in its turn depends on the firm's perception of possible approaches and of how the available expertise fits the requirements of each approach. The technology with which to realize each approach evolves over times and it is difficult for a firm to change its technological expertise. As a consequence each firm in the industry tends to perceive and select approaches which fit best with the existing technological expertise. A different selection of approaches by each firm, due to uncertainty, may thus be reinforced over time. This leads Cohen and Klepper to assume that when an industry consists of a large number of small firms, a greater number of the N approaches in the industry will be pursued than when the industry consists of a few large firms. The diversity of innovation approaches in an industry therefore increases with the number of firms in the industry. In the long run, high diversity may offer more innovations than low diversity. The existence of many small firms thus has a social advantage.

On the other hand large firm size may have social advantages too. Cohen and Klepper state that large firms usually are better able to appropriate their innovation profits.[20] One argument is that small and large firms will on average have the same cost of creating an innovation, while large firms have larger sales volumes. In the case of replacement of the 'old' product in a market, large firms are at an advantage. They may also be willing to undertake innovation projects with high technical or market uncertainty; not only because market uncertainty is relatively low, according to Cohen and Klepper's argument mentioned above, but also because the larger sales volume makes it easier to finance such a kind of project. Cohen and Klepper (1992, 4) relate to this when assuming that firms in an industry will spend an amount of money on R&D proportional to their sales in that industry. Not only does this imply an equal R&D intensity across firms of different size, but also a tendency of large firms to spend more per project, because the number of innovation approaches taken by a firm (M) does increase less than proportionally with firm size.

The larger amount of money spent by large firms on an innovation project may have two consequences. First, as argued by Cohen and Klepper, the type of innovation projects undertaken by large firms are on average more expensive and riskier than the projects of small firms. The societal advantage of large firms may therefore be that they undertake projects with large social benefits compared to private benefits. Second, and in conformity with the time–cost tradeoff curve, small and large firms may have similar projects, but large firms spend more in order to innovate faster.

We take a fictituous example to summarize the advantages of small and large innovating firms. Assume that a specific industry Z in countries A and B is of the same size; aggregate demand = aggegrate sales = 100. In country A the industry consists of two firms, in country B of ten firms. Per country the firms are completely the same in a number of respects. Table 5.3 gives an overview of these economic features of the firms.

If each firm has innovation projects different from all other firms in the industry then it follows from the data in Table 5.3 that in country A four innovation projects are undertaken, while in country B the 'technological diversity' is greater, because the total number of projects is ten. In contrast the smaller firms in B may not be able to begin projects like in A, because such projects need too many funds, or when they undertake them, they need to make R&D investments much slower in order to reduce total project costs by 60 percent.

Specific R&D expertise and rigidity

The assumption that each firm in an industry may have its own 'unique' choice of innovation projects also plays an important role in Henderson and Clarke (1990). While Cohen and Klepper (1992) concentrate on technological diversity among the firms in an industry at a specific moment in time, Henderson and Clarke focus on the change of the nature of the typical innovation project in an industry over time. This change may result in drastic shifts in market shares of incumbent firms and the exit and entry of firms to the industry. Table 5.4 presents data on market shares of the main firms in the photolithographic alignment equipment industry.

The first generation product in the industry was the contact aligner of which Cobilt was the market leader. The features of the next generation product, the proximity aligner, compared to the contact aligner are typical of the kind of technological change which occurred during the whole period:

> The introduction of the proximity aligner is clearly not a radical advance. The conceptual change involved was minor, and most proximity aligners can also be used as contact aligners. However, in a proximity aligner, a quite different set of relationships between componenets is critical to succesful performance . . . In particu-

lar, in a proximity aligner, the relationships between the gap-setting mechanism and the other components of the aligner are significantly different. (Henderson and Clarke, 1990, 24)

This kind of innovation is called 'architectural' innovation. Typical for this ideal type of innovation is that the overall technological concept remains the same, but the linkages between the components change. In contrast, a radical innovation is typified by a change of both the core concept and the linkages between the components. On the base of these two dimensions Henderson and Clarke distinguish four types of innovations, as Figure 5.3 shows.

An important difference between a radical and an architectural innovation is that the former is quite apparant for those informed about the industry as something drastically different from the existing product generation in the industry, while the latter looks very much like the existing product. For a radical innovation it is well known that they often stem either from an outsider of the industry (for example a small entrant) or from a new business unit of an incumbent firm. Architectural innovations by contrast have the same core technology as the old product and it is not immediately clear that product performance depends on new linkages between some components, which require quite a different technological expertise from the old product.

While learning new technological expertise is difficult, the problem for incumbent firms in the industry, confronted with an architectural innovation, is aggravated, because the way they perceive and learn about new technology is strongly influenced by the 'logic' of the old generation product. This means that it is difficult to fully understand all the technological details of the new logic. Table 5.4 suggests that the market leaders seem to have the greatest difficulties in following the innovator in the next stage. The explanation is based on the assumption that the market leader is the best in producing the old generation product and realizing a series of product improvements based on the same

Table 5.3 An industry with small and large firms

	Typical firm in country A	**Typical firm in country B**
Sales	50,0	10,0
R&D	5,0	1,0
R&D intensity	10,0 %	10,0 %
#projects	2	1
R&D/proj.	2,5	1,0

Table 5.4 The shifts in market shares in the photolithographic alignment equipment industry from 1962–86

Firm	**Alignment equipment**				
	Contact	Proximity	Scanners	Step and repeat 1	Step and repeat 2
Cobilt	44		<1		
Kasper	17	8		7	
Canon		67	21	9	
Perkin			78	10	<1
GCA				55	12
Nikon					70
Total	61	75	99	81	82

Source: Henderson and Clarke (1990, 24)

core technology and component technology interactions (interfaces). More generally Henderson and Clarke (1990, 14–15) state:

> With the emergence of a dominant design, which signals the general acceptance of a single architecture, firms cease to invest in learning about alternative configurations of the established set of components. New component knowledge becomes more valuable to a firm than new architectural because competition between designs [all based on the dominant design] revolves around refinements in particular components. Successful organizations therefore switch their limited attention from learning a little about many different possible designs to learning a great deal about the dominant design.

This specialization of learning about a single design is, according to Henderson and Clarke, reflected in several organizational aspects. They distinguish three aspects: communication channels, information filters, and problem solving strategies (ibid., 15–16). Communication channels (formal and informal) develop around those interactions within the organization which are critical to its task, which in the current context means the interactions between core technology and component technologies in accordance with the whole product architecture. The same holds for information filters which are needed to 'protect' the organization from the mass of irrelevant information from outside. People in the firm who have expertise about crucial linkages in the product architecture will gradually decide more and more which technological developments in the environment are important. Lastly, the engineers in the firm

Figure 5.3 The classification of innovations according to Henderson and Clarke

	Core concepts	
	Reinforced	Overturned
Unchanged	Incremental innovation	Modular innovation
Changed	Architectural innovation	Radical innovation

Linkages between core concepts and components

Source: Henderson and Clarke (1990)

will, during the series of product improvement of the dominant design, be confronted with recurring problems and design problem-solving mechanisms which are best suited. Channels, filters, and strategies are thus influenced by the product architecture and the more this is the case, the more difficult it becomes to detect and react to new architectural innovations in the market.

In this way Henderson and Clarke include organizational elements to explain shifts in market shares. The industrial organization approach would focus on the fact that strong competition in the photolithographic alignment equipment industry has brought about a series of new product generations, but would have had problems to explain why in a number of subsequent product innovations the market leaders turned out to be the losers in the next round. By focusing on specific types of innovation and related organizational characteristics Henderson and Clarke stress some evolutionary perspectives, such as path dependency (specialization, often leading to rigidities) in R&D and production.

5 CONCLUSIONS

In this chapter it became clear that the economic literature provides limited insights into how firms react to the innovation intentions and actual results of rivals. Empirical studies that are available on the topic do not conclude that an increase in the R&D or new products by rivals result automatically in an increase in one's own R&D. Economic theory has traditionally paid more attention to the degree of competition in general terms. One of the underlying reasons relates to the behavioral assumptions of firms. We discussed some of the important differences between neoclassical economic theory and evolutionary theory in this respect. An important difference between the two economic approaches with respect to innovation is the assumed ease or difficulty of imitation. More generally the way in which innovating firms are assumed to acquire new information and make decisions may have a considerable impact on how economists perceive the role of markets in innovation. This also has consequences for the perceived role of government. In the next chapter the 'traditional' roles of government in technology are discussed. Those roles are still largely based on the assumption that imitation is 'easy'. In Chapter 11 newer elements of technology policy are addressed.

NOTES

1. The nature of demand does, however, play an important role, though implicitly so, in the distinction we make between a market with homogeneous and a market with heterogeneous technology.
2. Foreign competition was measured as the import of high technology products in the US in the industries covered by the 308 firms. The importance of technologies in those industries was indicated by R&D intensity. Over the sample period 1972–86 the average R&D intensity for the 308 companies was 3.22 percent (Scherer, 1992, 137).
3. Note that we argued earlier that in most industries the knowledge underlying innovations is industry- or even firm-specific and that, as a result, imitation may be more difficult than (economic theory) assumed.
4. It concerns 10 major pharmaceutical firms in the U.S and Europe which cover 25-30 percent of sales and R&D in the total market (so not only antihypertensives).
5. They studied three industries: 8-bit microprocessors, class-8 diesel truck engines, and water treatment chemicals. The relation between process technology and concentration/growth rate was an 'inverse U-shape', while in the case of product technology a 'U-shape' was found. Although the latter is in contrast with a single life cycle of technology in the industry, it could indicate a second cycle, based on new product technology.
6. Each firm may thus create a new product or process, but each product or process innovation of a firm is quite similar to other innovations taking place in the market at the same time.
7. A patent is a legal protection against imitation; for more details, see Chapter 6.
8. On the other hand, each firm may require its knowledge base to develop in order to compete in the next round with the successful innovator. To the extent that a certain degree of technology competition is desirable from a societal point of view, there is no, or at least less, waste than we suggested. We come back to this point later.
9. More precisely: $Cs \leq nX$. For some or all other firms the innovation project may not be completed at the time the innovator gets the patent and brings its new product on the market, hence the R&D expenditure of a typical competitor j of innovating firm i is: $R\&D_j \leq X$.
10. The social benefit will differ, depending on whether there is a winner takes it all situation or not.
11. Theoretically, however, the issue is more complicated. While in the heterogeneous case diversity of approaches may lead to the optimal track of product technology, but the lack of competition may result in a suboptimal outcome for this optimal track, the homogeneous case may lead to a suboptimal track, but severe competition 'guarantees' an optimal outcome in this track.
12. This is the core question of the structure–conduct–performance approach in general. The degree of competition is related to various performance indicators of the firms in a market, such as

efficiency, profit margins, innovativeness, income distribution. (see, for example, Scherer and Ross, 1990).

13. The extreme case is a monopoly situation without any current or future threat of entry of another firm in the market.
14. Assuming that all competitors gain from economies of scale.
15. There is no guarantee that the new product is more advanced, because no one is punishing the monopolist for a bad innovation performance.
16. We exclude the possibility of a perfect patent for the first innovator in the market, because this would be a case of monopoly.
17. See, apart from the former section, also the brief historical picture drawn in Chapter 2.
18. See for example R.R. Nelson and S.D. Winter (1982).
19. Which, of course, is not true, but it simplifies the analysis.
20. In Teece (1986) it was argued that production and distribution facilities may be 'complementary assets' required for the appropriation of potential profits accruing from a new product.

REFERENCES

Acs, Z.J. and D.B. Audretsch (1987), Innovation, market structure, and firm size, *The Review of Economics and Statistics* 69, 4, 567–74

Acs, Z.J. and D.B. Audretsch (1988), Innovation in large and small firms: an empirical analysis, *American Economic Review* 78, 4, 678–90

Bain, J.S. (1956), *Barriers to New Competition*, Cambridge (MA), Harvard University Press

Baldwin, W.L. and J.T. Scott (1987), *Market Structure and Technological Change*, Chur, Switzerland, Harwood Academic Publishers

Bowman, D. and H. Gatignon (1995), Determinants of competitor response time to a new product introduction, *Journal of Marketing Research* 22, February, 42–53

Birnbaum-More, Ph.H., A.R. Weiss and R.W. Wright (1994), How do rivals compete: strategy, technology and tactics, *Research Policy* 23, 249–65

Cockburn, I. and R. Henderson (1994), Racing to invest? The dynamics of competition in ethical drug discovery, *Journal of Economics and Management Strategy* 3, 3, 481–519

Cohen, Wesley M. and Steven Klepper (1992), The tradeoff between firm size and diversity in the pursuit of technical progress, *Small Business Economics* 4, 1–14

Cohen, W.M., R.C. Levin and D.C. Mowery (1987), Firm size and R&D intensity: a re-examination, *Journal of Industrial Economics* 35, 4, 543–65

Gruber, W.H. and D.G. Marquis (eds) (1969), *Factors in the Transfer of Technology*, Cambridge (MA), MIT Press

Henderson, R. and K. Clarke (1990), Architectural innovation: the reconfiguration of existing product technologies and the failure of established firms, *Administrative Science Quarterly* 35, 9–30

Henderson R. and I. Cockburn (1996), Scale, scope, and spillovers: the determinants of research productivity in drug discovery, *Rand Journal of Economics* 27, 1, 32–59

Hundley, G., C.K. Jacobson and S.H. Park (1996), Effects of profitablility and liquidity on R&D intensity: Japanese and US companies compared, *Academy of Management Journal* 39, 6, 1659–74

Kamien, M.I. and N.L. Schwartz (1981), *Market Structure and Innovation*, Cambridge, Cambridge University Press

Kleinknecht, A. (1989), Firm size and innovation; observations in Dutch manufacturing industries, *Small Business Economics* 1, 215–22

Levin, R.C., W.M. Cohen and D.C. Mowery (1985), R&D appropriability, opportunity, and market structure: new evidence on some Schumpeterian hypotheses, *American Economic Review* 75, 2, 20–24

Levin, Richard C., Alvin K. Klevorick, Richard R. Nelson and Sidney D. Winter (1987), Appropriating the Returns from Industrial Research and Development, *Brookings Papers on Economic Activity* 3, 783–820

Mansfield, E. (1985), How rapidly does new industrial technology leak out? *The Journal of Industrial Economics* 24, 2, 217–23

Mansfield, E. (1995), *Innovation, Technology and the Economy*, Cheltenham, Edward Elgar

Meron A. and R.E. Caves (1991), Rivalry among firms in Research and Development outlays, paper presented at the UBC Summer IO Conference

Mowery, David and Nathan Rosenberg (1989), *Technology and the Pursuit of Economic Growth*, Cambridge, Cambridge University Press

Nelson, R.R. and S.D. Winter (1982), *An Evolutionary Theory of Economic Change*, Cambridge (MA), Belknap Press

Pavitt, K. (1982), R&D, patenting and innovative activities, *Research Policy* 11, 33–51

Pavitt, K. (1984), Sectoral patterns of technical change: towards a taxonomy and a theory, *Research Policy* 13, 343–73

Reinganum, J.F. (1989), The timing of innovation: research, development and diffusion, in R. Schmalensee and R. Willig (eds), *Handbook of Industrial Organization*, Vol. 1, Amsterdam, North Holland, Chapter 14, 850–905

Scherer, F.M. (1992), *International High-technology Competition*, Cambridge (MA), Harvard University Press

Scherer, F.M. and D. Ross (1990), *Industrial Market Structure and Economic Performance*, third edition, Boston, Houghton Mifflin

Teece, D.J. (1986), Profiting from technological innovation, *Research Policy* 15, 286–305

6. The role of government in stimulating innovation[1]

1 INTRODUCTION

In the foregoing it has become clear that firms, when confronted with competitors in their markets, do not automatically solve all the problems with respect to innovation. Or put in terms of orthodox economic theory: markets do not function optimally with regard to the creation of innovations. Firms are confronted with technological and market uncertainty and, from the welfare perspective of neoclassical economics, it can be said that markets have strong imperfections when the 'trade' of knowledge is concerned. In this chapter it is discussed how economists perceive the role of government to solve part of the abovementioned problems. Traditionally this role focuses on two points: the use of patents, and the provision of R&D subsidies and taxes. We limit this chapter to these two facets of government involvement in stimulating innovation. In Chapter 11 we discuss some new elements of modern technology policy, which have not been included in the traditional economic analysis of innovation. These modern aspects not only influence policies to stimulate the creation of innovations, but also policies to stimulate the diffusion and exploitation of (new) technology.

We begin by describing how the patent system operates in practice. Thereafter the theoretical arguments for having a patent system are given. The second part of the chapter is devoted to subsidies and taxes.

The problem

Innovation has been defined as the process through which new products or processes are succesfully applied. Chapters 1 and 3 made clear that R&D is only one of a number of different activities required to develop the new product or process and exploit it successfully in economic terms. Traditionally technology policy focuses on R&D as the main activity for successful innovation. In that sense it is rather limited. In this chapter we focus on the narrow view of innovation and technology policy. In Chapter 11 a broader view is discussed.

Private innovators are concerned with receiving a net benefit from their R&D efforts.[2] Normally they will therefore not undertake innovation projects without a sufficiently high net benefit. From a societal point of view the creation of a maximum number of 'good' innovations that are used as fully as possible in the economy is desirable. In Chapter 4 we already gave some examples of projects without a net private benefit but with considerable net social benefits.

The problem posed in theoretical terms is that the level of R&D that maximizes the net benefits of private firms is not equal to the R&D level that maximizes society's creation and full use of new technology. In most cases net private benefits of an innovation are smaller than its social benefits. Moreover private firms try to limit the use of their innovation by others if that use means a reduction of their own benefits from the innovation. This conflict between private firm and state is mitigated, because the nature of their R&D activity may differ: private firms favor small projects or large projects with low uncertainty, while government focuses on large and highly uncertain projects that, when they fail, endanger the continuity of the firm.

As we have seen, the firm's uncertainty has a technological or a market origin. Technological uncertainty is related to the complexities of the relevant technology(ies), the life cycle of these technologies and technological opportunities provided by the general technological and scientific developments in the field(s). Market uncertainty is about how competitors react to your R&D expenditure or innovation results, or more generally, about a firm's uncertainty about the nature and introduction time of new technology.

According to Chapter 4 uncertainty might lead to underinvestment in R&D and a level of innovative activity that is undesirable from a societal point of view. Innovators need to be stimulated. One way is to reduce market uncertainty by providing a legal protection against imitation. That is what a patent is about. But what is the 'optimal amount of protection' from a society's point of view? When a firm's innovation is protected from imitation the use of that innovation is limited. How can innovators be stimulated to undertake more R&D and how is the diffusion of new knowledge in an economy at the same time facilitated? In theory the patent system provides a solution. We will see, however, that some doubt exists as to whether this solution is efficient in all cases.

Another way to stimulate innovation by private firms is to provide subsidies or favorable tax schemes for R&D. An important aspect of technological uncertainty is the total costs of an innovation project. Through subsidies or taxes the R&D costs for the firm will be reduced and, indirectly, technological uncertainty will be diminished.

2   THE EMPIRICAL SIDE OF PATENTS

In this section it is first explained what a patent is and how an innovator can get one. Thereafter we will explore what kind of firms get patents and for which kinds of new technology. The section concludes with an examination of the 'fit' between innovation activity and patents. An important question that arises is to what extent patents are actually giving the protection suggested.

The patent system

A patent is the exclusive right provided by the government to commercialize a new invention during a limited period of time. The owner of such a property right is thus entitled to exclude other economic agents from fabricating, using or trading this invention. In most countries, including European countries, the exclusive right is according to the national patent law provided to the person or institution first filing an application for a patent to a specific invention. In contrast, the US patent law sees the actual first inventor as the rightful claimant for the patent.

We abstract from this difference and will use the term inventor as the person or institution applying for a patent. That is not to say that every inventor applies for a patent; this matter is discussed in the last paragraph of this empirical section.

Applications for a patent are filed at the patent office of the country where the inventor has residence. In most cases the inventor is represented by a patent agent who is officially registered by the government as a specialist in the field of patenting. The inventor hires the expertise of the agent who does the paperwork that needs to be sent to the patent office. The information concerning the patent is studied by scientists and engineers of the patent office who decide, when necessary with the advice of outside specialists, whether a patent can be granted. The patentability of an invention depends on several things which may differ among countries. The most common and important criteria are 'newness' and 'usefulness' of the invention (Kaufer, 1989). The inventor is required to reveal enough information about the invention for the patent office to make a proper judgment about newness and usefulness. The patent agent knows what kind of information is requested by the patent office and he is responsible for receiving this information from the inventor. The information thus disclosed by the inventor is public information and it is supposed to enable a 'normal' specialist in the field to replicate the invention.

National patent offices receive numerous applications each year that must be carefully studied. The time between an application for a specific invention and the decision by the office as to whether a patent can be granted or not, is considerable. It takes approximately two years on average to administrate a patent application. When a patent is granted the inventor gets the right to exploit the invention commercially for a certain period. In Europe most countries use a period of 20 years. The inventor has to pay a fee when the patent is granted.[3]

In many countries, once the patent is granted the applicant does not automatically receive the property right for its invention for the full legal period. Countries such as Germany, Austria, France, Great Britain, the Netherlands, Japan, Switzerland and the United States require periodic renewal fees to be paid to maintain the patent rights (Kaufer, ibid.,13). Alternatively inventors can apply for a so-called 'petty patent' in Germany, Japan, Italy, Spain, and several other countries (in Germany this is known as 'Gebrauchsmuster'). This petty patent is for a shorter period and provides less rights than a full patent, but the costs are less and the time between application and granting of the patent is much shorter. From an economic point of view it is interesting to know under which circumstances one applies for a full patent and when for a petty patent. And also, when and why do inventors stop paying renewal fees? There are important differences between countries in the kind of inventions that can be patented. For example, Switzerland used to exclude chemical substances, and until 1978 Italy prevented drugs from being patented.

Quite a different matter is the geographical area for which the inventor may get a patent. As already suggested in the foregoing, normal practice has been to receive a patent for your own country. The patent therefore provides protection against imitation from domestic firms, and domestically operating foreign firms. The more international the market is – and this is the case for an increasing number of markets – the more the inventor is in need of a patent which covers his whole economic market, not his administrative region of residence. During the last three decades, an increasing number of patents have been granted for several countries, apparently the countries which are the main markets for the inventor's new product. Well known is that many firms try to get a patent for the American market in addition to the patent for their home country. Within Europe, inventors used to apply for a patent in every country they traded with. Obviously a more efficient solution was needed.[4]

Today an inventor in one of the member states of the European Union can receive a patent for the whole union by applying once to the European Patent Office in Munich.

Table 6.1 presents, among others, data about the growth of the number of patent applications from 1980 to 1990 and the ratio of external applications and applications by residents in OECD countries.

Who are the inventors and what do they patent?

As we have seen in Chapter 3, innovation activities in especially the larger firms have been formalized during this century as R&D activities in a separate department. As we will see this trend also reflects in the number of patents owned by firms and the number owned by individual inventors. Originally inventions were the domain of individuals and until the early 1960s

Table 6.1 Number of patent applications in OECD countries

Countries [1]	Number patent families [3]	As a % of OECD	Growth in PF 1990/80 (%)	External appl [2]. Resident appl.
USA	22164	29.3	60	0.26
Japan	16318	21.6	85	0.11
Germany	13482	17.8	12	0.36
France	5402	7.1	26	0.33
UK	4497	6.0	24	0.40
Italy	2606	3.5	44	0.30
Switzerland	2372	3.1	-10	0.40
Netherlands	2267	3.0	43	0.34
Sweden	1330	1.8	-7	0.41
Austria	867	1.2	-4	0.56
Canada	827	1.1	57	
Finland	703	0.9	92	0.53
Australia	694	0.9	61	
Belgium	616	0.8	95	0.36
Denmark	465	0.6	76	0.40
Spain	399	0.5	51	0.10
Norway	251	0.3	25	0.23
Ireland	227	0.3	149	
Luxembourg	79	0.1	-25	0.54
New Zealand	39		-43	
Greece	25		39	
Portugal	3			
Iceland	3			
W. Europe	35594	47.1	17	0.37
EC	30068	39.8	21	0.36
EFTA	5526	7.3	0	0.49
N. America	22991	30.4	60	0.26
Japan	16318	21.6	85	0.11
Oceania	733	1.0	47	
Total OECD	75636	100.0	40	

[1]   In the case of non-European countries (USA, Japan, Canada, Australia and New Zealand), data cover all intentions for which protection has been sought in at least one western European country. For EC and EFTA countries they cover all those for which protection has been sought in at least one other European country. The country of origin cited is that of residence of the filer.

[2]   External applications for a given country correspond to the number of application by its residents for patents in the other trade zone in which it most often seeks protection. Resident applications cover all patent applications by resident of a country regardless of whether protection is sought abroad. The ratios for groups of countries are based on the national data available

[3]   The number of patents are grouped in patent families (PF)

Source: OECD (1994, 192)

a majority of the radical inventions reported by Jewkes et al. (1969) were created by individuals or small firms. More recent empirical studies show a more prominent role of large firms. According to Kaufer (1989, 16) US corporations received only 7 percent of all patents issued in the United States, while 82 percent went to unaffiliated individuals. By the early 1980s the share of American individuals had fallen to 18 percent and the share of domestic corporations rose to 42 percent. Meanwhile the share of foreign corporations had risen to 30 percent.

We have already seen that in some periods and some countries particular products or other inventions have been excluded from patenting. What kind of new technology do inventors patent? This seems to be a somewhat neglected area in empirical studies on patents and patenting activity. We are especially interested in how patented inventions relate to new products and new processes that are being introduced in the economy. If a single patent or a group of related patents cover a complete product then, at least theoretically, monopoly profits can be gained during the legal period of protection. In many cases a patent is applied for before the invention has been developed into an innovation. Only afterwards, when the innovation is commercially applied, it becomes clear whether one or more patents on the new technological knowledge embodied in the innovation are profitable.

The number of patents covering a new product or process differs widely, depending among other things on the complexity of the new technology and the characteristics of each field of technology. It is clear that the patent office, representing the interest of the country, must strive for 'narrow' patents to reduce the extent of monopoly. In contrast innovating firms and individuals may want to 'maximize' the fields of application covered by the patent.

At this point we need to clarify a particular issue. In the introductory chapter, invention has been defined as the outcome of the applied research stage, that is, as the prototype. From what has been revealed about the kinds of new technology that are patented it is clear that inventions, in the above sense, are only a limited portion of the number of patents. Pavitt (1982), among others, has stated that patents are applied for to protect the innovation, not the invention. Thus not only the invention (that is, prototype) will, if necessary, be patented, but every new technological element that came up during the period of idea until test production and marketing. The word invention will be used for any new technological item within the whole innovation process.

Another point is, how much effort inventors are willing to make to get a patent and to make the legal protection against imitation as effective as possible? Inventions will normally differ in value for the inventor. For a full patent he pays a fixed fee, no matter what the anticipated value of the invention is. We saw, however, that some countries provide the possibility of a petty patent and in other cases, sometimes in addition to the fee for a normal patent, renewal fees have to be paid. The way in which inventors behave in their patenting activity presumably relates to the value they perceive of the invention being protected. The number of systematic studies about this behavior is very limited. According to Schankerman and Pakes (1986) the bulk of patents are cancelled, despite rather low renewal fees. Only a small minority of the total number of patents carry through the whole legal period. This suggests that at least these patents have a high value to the inventor. One of the reasons to cancel a patent is that no innovation came out of it. Another reason is that the inventor, maybe after the patented prototype has been developed into an innovation, discovered that the economic value of the patent is much lower than he originally believed.

The use of patents and innovation activity

In the previous section we were concerned with the things that are being patented. The question is here how many inventions are patented? Do all inventions relate to innovation? Do patents provide the protection against imitation, as the original idea behind the patent system suggested? These questions give us an important lead to whether patents are a good 'output indicator' for innovation activity.

The picture that can be derived from empirical research is that many inventions are not patented. An important question is which proportion of inventions are patented and whether this proportion varies from country to country, over time, between industries, and between different types of firm (for example small and large firm, monoproduct versus diversified firm). This proportion is called 'the propensity to patent'. One of the most extensive studies about differences in the propensity to patent is done by Scherer (1983). He found that the number of patents received in 1976–77 by 443 US (large) corporations per million dollars of 1974 company-financed R&D expenditure ranged from 0.45 to 3.98 (the industry mean was 1.70). More detailed analysis learned that the propensity of patenting was strongly influenced (80 percent of the variance explained) by both the amount of R&D expenditure related to the product line for which a patent was granted and by the particular industry to which the line belonged (Scherer, 1983). Comparison of patenting activity of different industries with the aim to compare innovation peformance must therefore be done with some care.

Another important finding is that many patented inventions do not carry through to commercially successful innovations. The question is why? Are some inventions patented while it turned out later that the project failed? Or did a competitor introduce its new product earlier? Are some patents obtained for other reasons than protection of a specific innovation? Figure 6.1 gives an impression of the number of innovations, of patents, and of inventions.

Figure 6.1 Inventions, patents, and innovations

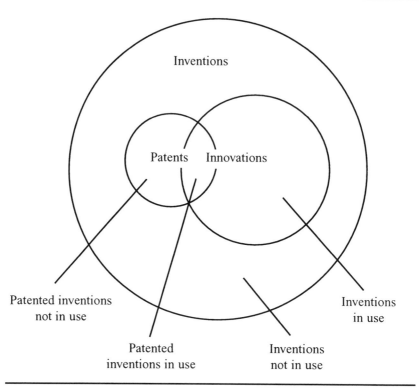

Source: Basberg (1987, 133)

While the propensity to patent varies by industry it may also vary by country and over time. 'Several authors have noted an apparent world-wide decline in patenting activity during the 1970s'. (Schankerman and Pakes, 1986, 1071) This led to questions about a decrease in R&D productivity that might explain the productivity slowdown (especially in the US) in the same period.

The last and most fundamental question we want to explore in this section is whether patents provide 'complete' protection against imitation. And related to this, do they allow for monopoly profits that form the stimulus for antici-pating firms to invest in uncertain R&D?

It will not come as a big surprise that patents do not always provide full protection. The more relevant question is when and to what extent do they pro-vide protection? And, related to this, do inventors have alternatives?

A number of these questions can be answered on the basis of an empirical study carried out by Levin et al. (1987) in the US (for a European study, see

Taylor and Sibbertson, 1973). In this research the importance of patents compared to other means of appropriating innovation profits is examined. Managers of over 500 (big) US firms were asked how important patents are to protect their innovations. Only in a limited number of industries are patents perceived as an important form of protection, as Table 6.2 shows.

This does not imply that in most industries the whole issue of protecting innovations against imitation is unimportant. In pratice other forms of protection exist. One way is to keep the innovation secret. Table 6.3 reveals that having a 'lead time' over competitors and the closely related concept of 'learning curve' are the most important mechanisms to appropriate innovation profits.

Table 6.2 The difference in importance of patents by sector

Industry	Process patents		Product patents	
	Mean	Standard error	Mean	Standard error
Pulp, paper, and paperboard	2.6	0.3	3.3	0.4
Cosmetics	2.9	0.3	4.1	0.4
Inorganic chemicals	4.6	0.4	5.2	0.3
Organic chemicals	4.1	0.3	6.1	0.2
Drugs	4.9	0.3	6.5	0.1
Plastic materials	4.6	0.3	5.4	0.3
Plastic products	3.2	0.3	4.9	0.3
Petroleum refining	4.9	0.4	4.3	0.4
Steel mill products	3.5	0.7	5.1	0.6
Pumps and pumping equipment	3.2	0.4	4.4	0.5
Motors, generators, and controls	2.7	0.3	3.5	0.5
Computers	3.3	0.4	3.4	0.4
Communications equipment	3.1	0.3	3.6	0.3
Semiconductors	3.2	0.4	4.5	0.4
Motor vehicle parts	3.7	0.4	4.5	0.4
Aircraft and parts	3.1	0.5	3.8	0.4
Measuring devices	3.6	0.3	3.9	0.3
Medical instruments	3.2	0.4	4.7	0.4
Full sample	3.5	0.06	4.3	0.07

Source: Levin et al. (1987)

Table 6.3 The value of patents and other forms of protection against imitation

Method of appropriation	Overall sample means		Distribution of industry means	
	Processes	Products	Processes	Products
Patents to prevent duplication	3.52 (0.06)	4.33 (0.07)	2.6–4.0	3.0–5.0
Patents to secure royalty income	3.31 (0.06)	3.75 (0.07)	2.3–4.0	2.7–4.8
Secrecy	4.31 (0.07)	3.57 (0.06)	3.3–5.0	2.7–4.1
Lead time	5.11 (0.05)	5.41 (0.05)	4.3–5.9	4.8–6.0
Moving quickly down the learning curve	5.02 (0.05)	5.09 (0.05)	4.5–5.7	4.4–5.8
Sales or service efforts	4.55 (0.07)	5.59 (0.05)	3.7–5.5	5.0–6.1

Source: Levin et al. (1987)

The second point we mentioned can now be phrased as 'if patents provide strong protection does it follow that monopoly profits can be gained?' In general, the answer is again 'no'. This relates to the fact that many patents do not cover the full range of technologies with which firms in a market compete and because technology is only one dimension of competition. The range of applications covered by a single patent is called the 'scope' of the patent. Lerner (1994) studied the scope of patents in the biotechnology industry. As a proxy for patent scope he used the number of subclasses of the International Patent Classification the patent was assigned to by the examiners of the patent office. A sample of 1678 patents from 173 private biotechnology firms was constructed and 553 rounds of financing by these firms were studied. It turned out that the value of the biotechnology firms is positively influenced by the scope of the patent(s) assessed by the venture capitalists.

Patents do usually not provide considerable monopoly power. When they do, it is because the scope of the patent is relatively wide. Important aspects of this limited degree of protection by patents have already been discussed in Teece (1986). Only in combination with already existing market power, especially (other) entry barriers, does a patent granted to a firm provide a ground for monopoly profits (Kaufer 1989, 40, 41).

3 THEORETICAL ISSUES ON PATENTS

Patents are one of the most extensively discussed topics in the theoretical literature on innovation. It is a delicate task to do justice to the numerous publications on the various economic aspects of patents. As argued earlier we concentrate on topics that bear a direct relation to the empirical characteristics and trends discussed in the previous section. There it came about that only in a few industries are patents seen as the best way to appropriate the benefits of an innovation. Clearly patents are an imperfect appropriation mechanism and we will look for theoretical discussions which embody this empirical notion. In our discussion of this topic and of some additional more theoretical topics we make a connection with our earlier used distinction between homogeneous and heterogeneous technology competition (see Chapter 5).

Theoretically the patent system solves the tension that exists between private and public interests in R&D and innovation. The tension can be described as follows. From the private point of view individual firms are stimulated to undertake R&D when the profits from successful innovation projects can be fully appropriated. From the public point of view, however, full disclosure of R&D results of all firms gives the best results, collectively. A patent provides legal protection for a limited period and thus a stimulus for the firms. At the same time the technological knowledge for which the patent was granted becomes public information. Thus, in return for a full disclosure of the new technological informaton of the innovation, the patent system provides a monopoly, although limited in various aspects. In the literature this connection between patenting behavior and the expected economic value of related new products or processes is first stated by Schmookler (1966). In his so-called demand inducement hypothesis it was stated that the returns to an invention vary directly with the number of units of output that embody the invention, or the size of the market.

One must bear in mind that many inventions are patented before their full economic value becomes clear. The propensity to patent can therefore not be following exclusively from the estimated net benefit of the innovation(s) derived from the invention. Rosenberg (1976) argued that technological opportunities play an important role in the patenting behaviour of firms. In the short run each market has a limited pool of technological opportunities. When this pool is getting 'empty' less patents will be applied for by the firms. In fact the argument for the decline in patent activity in the 1970s was the exhaustion of technological opportunities in major industries (Schankerman and Pakes, 1986). Only new scientific discoveries and fundamental/generic research may replenish the pool and create new opportunities for applied research and thus for patents.

There may be considerable differences between companies and sectors in the kind of 'intellectual goods' for which patents are granted. For example, in the pharmaceutical industry one or a number of patents may apply to a whole new product, a so-called 'new chemical entity' (NCE). This NCE can be seen as the prototype in the simple R&D model, as the result of the applied research stage. This patented new prototype is further developed and tested. In the electronics industry, by contrast, many innovations and patents cannot be clearly related to a single new prototype. As has already been stated in the empirical section, protection of the innovation is the main goal of patents. Patents on inventions, such as a new prototype, can therefore only be part of the total set of patents. Patents exist in every stage of the R&D process, including those activities which are sometimes not included in the definition of R&D (that is, engineering, design). This would imply that firms not only have innovations without official R&D (see Napolitano, 1991), but also patents.

Patents are a way to protect the results of a firm's innovation projects from imitation by competitors (or other firms). There are other ways of protection. Large firms sometimes undertake more R&D to construct an entry barrier in the market and as a protection against imitation. By providing a continuous stream of (minor) innovations they keep ahead of imitators (lead start, lead time). This issue of market power has been discussed in Chapter 5.

Levin et al. (1987) showed results of the largest empirical study on the importance of patents and other forms of protection. These and other empirical studies reveal that the importance of patents as a form of protection of innovation profits is limited. If other forms of protection are more important than patents, does that mean that imitation and the diffusion of knowledge among firms in the economy are less a problem than the theory of patents suggests? In Chapter 5 the costs and time needed to imitate were presented. This relates to the question of how exactly new know-how is acquired; we come back to that later on.

Inperfect appropriability

Orthodox economic theory assumes a high probability of imitation of the innovating firm's innovation. Empirical studies on imitation reveal that the cost of imitation may differ considerably between innovations, ranging from 25 percent to over 100 percent of the orginal innovation costs (see Table 5.3) The patent system offers the same period of protection, no matter what the perceived probability of imitation is. When a new product is brought on the market for which a patent exists, there may be no monopoly profits at all. This is the case when one or more competitors bring a superior new product on the market that is not violating the patent. The evolutionary theory of economic change (Nelson and Winter, 1982) emphasizes this dynamic aspect of competition, the continuous introduction of different new products in one and the same market. Within this context patents offer only one out of a number of ways to protect the innovation profits. In the chemical and pharmaceutical industry patents do provide an effective means of protection. But even there it is possible that various new chemical entities, each of which has a waterproof patent, can be used to cure the same disease.

In sectors such as the electronics industry, the probability that a new product creates a (temporary) monopoly is much lower, because products based on different technologies compete against each other, or a new product consists of a number of new subsystems which are not all covered by patents in the hand of a single firm.

When other forms of protection of intellectual property are more important than the patent, does that imply that imitation and the rapid diffusion of knowledge among firms in the economy are less problematic for the innovating firm than the theory underlying the patent system suggests? This appears to be the case; imitation can be costly and the time required to follow gives the innovator enough of a lead.

The foregoing makes clear that a patent may have quite a different meaning for a homogeneous market than for a heterogeneous market. In a homogeneous market imitation is relatively easy (high speed, low costs) and more generally the diffusion of knowledge created and integrated within the whole innovation activity from the innovator to others is not problematic.

By contrast, in a heterogeneous market imitation and diffusion are costly and complex, which implies that patents may not be needed to protect the innovation profits. Given the costs of patenting and the fact that the information, on the base of which a patent will be granted, is public, firms may decide to look for alternative forms of protection.

To the extent that mainly firms in homogeneous markets use patents as a means of protecting the new technological knowledge of their innovations, all that has been said about the costs and benefits of patents must be viewed accordingly.

In the empirical section it was argued that the value of the innovation after the patent(s) and compared to the situation without a patent will determine patent behaviour of the firm.[5] There is some room for a more subtle approach. The situation without a patent can be compared to various constellations with a patent, namely a patent for the full length of the legal period, various shorter periods, depending on renewal fees which are actually paid for, and minor patents (petty patents, and the like).

In the theoretical discussions patent time is the main topic related to the above considerations. We will make a distinction between theories for homogeneous markets and theories for heterogeneous markets.

Optimal patent time

A large part of the theoretical discussions on patents focuses on the optimal period under which legal protection is provided. Most of these theoretical models of patent protection assume, although implicitly, a homogeneous market. The models are quite similar to the ones discussed in Chapter 5 on technology competition. Well known in this respect are the models of Nordhaus (1969, 1972). In these models a tradeoff is reflected between the value of the innovation for society and the welfare loss of a monopoly. The higher the societal value, for example the higher the unit production cost reduction brought about by a process innovation, the longer the period of patent protection should be. The higher the welfare loss due to the monopoly created by the patent, the shorter the optimal patent time for society.

In the light of what we know from practice the models discussed in the literature to which we referred above, have a number of drawbacks. First, many of them assume that a patent provides perfect appropriability conditions. This is less plausible the more the situation can be typified as market heterogeneity. Second, in most cases the theoretical model of pure competition is applied. If the market is characterized by a limited degree of competition (oligopoly, monopolistic competition), the analysis becomes more complicated because interactions between firms have to be brought in, and conclusions concerning optimal patent times may change drastically. Third, markets are assumed to be homogeneous in many cases, and in others no explicit distinction is made between homogeneous and heterogeneous markets. We have seen that the implications of the two are quite different. An example is the societal cost of duplication, sometimes incorporated in the models. In a homogeneous market duplication of R&D by the firms involved in a patent race may cause considerable societal costs. In a heterogenous market, however, 'duplication' means that some or all firms are trying to come up with a new generation product (or process), and by doing so they create technological diversity.[6] Fourth, innovations are assumed to have no influence on each other. In Chapter 5 it was shown that evolutionary economics distinguishes markets with discrete technological change from markets where the technological change is cumulative. In markets with cumulative technological change the value of an innovation has two sides: the effect in terms of cost reduction or product improvement (for process and product innovation respectively), and the effect on a number of subsequent innovations.

Apart from arguments against the theoretical treatment of optimal patent time, there is also an empirical argument not to put to much emphasis on this issue. In the empirical section we saw that many patents are not renewed. This fact led Lerner (1994, 319) to the statement that the emphasis shifted from length to breadth of a patent: 'But as empirical work has made clear that the effective protection provided by a patent is often less than its legal life (Mansfield, 1984), several theoretical articles have turned to examining the optimal breadth of patent claims'. We first discuss the more general topic of diversity and then turn to the breadth or scope of the patent.

Patents and technological diversity

Seen from the perspective of homogeneous markets a number of firms, all participating in a race for a patent, may invest more in R&D than what is optimal for society. There are basically two merits in this situation. First, technical progress may be accelerated. Second, the probability that one of the firms indeed solves all of the technical problems and also brings the new product on the markets successfully is higher the larger the number of firms. Once diversity is introduced, the picture changes:

> conducting multiple R&D projects in a given field of technology not only accelerates progress, but leads to a diversity of inventions and technological solutions, each of which, especially in the case of product inventions, has advantages in satisfying the diverse tastes of different consumers. Models examining the question of optimal product variety show that it is possible under plausible conditions for there to be too few competitive approaches, too many, or just the right amount, with no guarantee that the third result will dominate . . . That the results of these relatively narrowly-focused analyses can be so ambiguous is in itself disconcerting. In addition, little progress has been made to date in integrating the work on optimal product variety with the theories of optimal patent life and strength. Thus, much remains to be learned. (Kaufer, 1989, 40)

Another issue, more recently discussed, is patent scope. The scope of the patent is the size of the technological or functional area of application covered by the patent. The allowable breadth of patent claims is determined by the patent examiners and the judiciary (Scotchmer, 1991, 30). What is the optimal breadth of a patent? As with patent time each invention may have a different optimal scope, depending on its economic value. Similar to the studies of Nordhaus mentioned before, several economists examined scope by looking at its two main consequences: the incentive to invest in R&D and the welfare loss due to monopoly (see for example Gilbert and Shapiro, 1990).

However, the value of a patent also depends on the number of innovations derived from the invention that are not covered by the patent. We have seen that some innovation projects are highly uncertain because outcomes may have a wide range of applications for which the firm does not have the production and marketing knowledge. The incentive for innovative firms to invest in such projects is reduced when the authorities only grant narrow patents. On the other hand when patents are wide other firms are discouraged to under-take R&D within the 'trajectory' controlled by the original innovator. As a consequence diversity in the number of firms in the industry involved in a technological trajectory (similar to Cohen and Levinthal in Chapter 5) is in the limit reduced to a single firm. Merges and Nelson (1994) conclude that the involvement of a diversity of firms in every technological domain of innova-tion provides the best result for the whole economy. This would mean that patent scope should be rather restrictive. According to them patent offices have not been consistent in this matter. The main reason no doubt is that from a practical point of view it is very difficult to apply criteria for scope.

Conclusions with respect to patents

One of the main problems innovators are confronted with is the uncertainty of R&D investment. The uncertainty about solving the technological problems is one aspect, being confronted with the possibility that competitors imitate, once you have solved all technical problems, is another. Patents offer legal protection to the innovator against imitation. By providing a patent and getting full disclosure of the patent information, the government seeks a balance between the stimulating effect of a potential monopoly position for the inno-vating firm and the disadvantage of actually establishing a monopoly for a single firm.

Theoretically a patent reduces market uncertainty for the innovating firm. As a result investment in R&D may rise. Whether this actually happens, and if this is a good thing for society, remains to be seen. The main disadvantage of the patent system is that the length of the period of protection is equal for all kinds of innovation. In theory a firm should pay a fee that is related to the eco-nomic value of the patent in return for monopoly profits. In practice, patents seldom create a monopoly situation for the innovator. An alternative is to have flexible patent lives. In practice, however, it is difficult and costly to estimate the optimal patent time. The issue of patent scope complicates matters even more.

The theoretical complexity of assessing the cost and benefit of the patent system can be reduced by the use of the distinction between homogeneous and heterogeneous markets. For a homogeneous industry a given patent time may have quite different consequences than for a heterogeneous market. An impor-tant consideration is the use of alternative forms of protection against imita-tion. It became clear that managers in many industries perceive patents as less important than lead time, secrecy, or learning curve. If new empirical studies support this outcome, the economic importance of patents will be seriously reduced.

The trade-off between the monopoly provided by the patent and the full disclosure of patent information gives a clue about the problematic character of information dissemination. Apparently government authorities argue that the dissemination of technological information in society is suboptimal. Some theorists explain this in terms of inefficiency of the knowledge market. Because knowledge has properties of a public good, of indivisibility and because it creates considerable (positive) externalities, the trade of 'pieces of knowledge' under a regime of strong competition is hampered. In more practical terms: the value of a piece of information that a firm wants to use as an input to its innovation process depends heavily on an exclusive full disclosure of its properties to the innovating firm and not to other firms. There can thus never be a 'bidding process' to establish the value of the information. In addition, full appropriability of the piece of information is difficult as it can be easily reproduced by the supplier (assuming that most pieces of information cannot be patented).

4   INTRODUCTION TO TECHNOLOGY POLICY

In the first part of the chapter the role of patents has been described. The patent system is not the only direct way in which government tries to influence technological developments brought forward by private firms. In this part of the chapter the main ways in which governments of western industrialized countries used to intervene in the innovation processes within markets are examined. As mentioned earlier, R&D subsidies and R&D tax schemes are the main elements of what is normally called 'technology policy'. We also discuss how national government stimulates fundamental research. The relations between fundamental research (science) and technology (innovations) have been discussed in Chapter 3. The main conclusion was that the picture of science providing inputs to the innovation processes of private firms is too simple. In particular in high-technology fields the impact of innovation results on science is as great as the other way around. The interactions between science and technology are part of the system(s) of innovation discussed in Chapter 11.

First, an overview is given of the R&D subsidies and taxes in Europe. The second section is devoted to a theoretical discussion of these instruments used by government to stimulate innovation.

The problem of technology policy

Today government in many western countries undertakes measures to stimulate innovation and the application of new technology in private business. The main question is whether this involvement of government indeed has the positive effects on the economy that one expects. The problem is basically twofold. First, by undertaking these measures government is influencing decisions of private business and one could argue that this 'market intervention' may reduce the ability of managers and others in private firms to react properly on future market developments. Second, while most of the measures cost money the question is whether their social benefits are greater than their social costs. These costs are to a certain extent straightforward (the costs of the government measures), but the benefits may depend on what other governments do. If one country stimulates innovation by its private firms another may be tempted to do the same, because otherwise its firms may, at least in the short run,[7] lose their competitiveness. This is what one calls the matching argument; one country matches the other with respect to measures to stimulate innovation.

As we have already seen in the sections about patents, new knowledge is basic to innovation. The main theoretical problem to private investment in R&D is that markets for knowledge related to technological innovation do not work very well. One solution is to undertake R&D in public or publicly-financed research institutes. In fact a large part of fundamental research is organized in this way. This came about in our discussion of main problems at the micro level and technology competition. Another solution is to reduce the barriers which prevent knowledge markets from functioning efficiently or to 'compensate' for such inefficiencies. It is a delicate task to provide a theoretical explanation for governments' involvement in innovation. As it turns out that most governments are involved, we need to discuss the theoretical arguments and try to understand the why and the how. The picture is very complex and we limit ourselves in this chapter to governments' efforts to promote the R&D of private firms. These efforts are to some extent in line with the neoclassical view on markets which work imperfectly with respect to innovation. In the second part of the book we deal with other kinds of government involvement, such as the promotion of diffusion of knowledge from the technological infrastructure of the country towards the firms.

5   THE EMPIRICAL SIDE OF SCIENCE AND TECHNOLOGY POLICY

In this section we briefly discuss science policy and present a more extensive picture of the most important ways in which governments of European countries are involved in their firms' R&D efforts. The role of the European Community will be discussed in the last chapter of the book.

Each industrialized country's government is involved in a series of activities and policies that are intended to create a 'prosperous climate' to do business. In that way it also stimulates innovation, because firms will on average have better financial positions to undertake R&D than in situations without such a climate. We will not discuss these government policies to create a favorable economic climate in general. We will focus on policies which are directly aimed at enhancing innovation. These policies are normally included in science and technology policies.

Science policy is concerned with higher education and with public research. Technology policy focuses on measures to stimulate innovation in private firms directly.

Science policy

A large part of what has been called fundamental research in Chapter 1 is undertaken or financed by national government. It was argued there that fundamental research does not have a direct commercial impact, in the sense that no 'fundamental' knowledge is directly applied in new products or processes. Indirectly, however, results from fundamental research and science in general stimulate innovation projects in which knowledge finds practical application. On the one hand the universities educate people who will find jobs in the laboratories of private firms. On the other hand research findings in universities and other public institutes enable applied research and development in private firms. In short, there are spillovers from science to all kinds of technological applications. Making patent information public is one element in governments' strategy to make science and technology as widely available as possible. And because of those spillovers private firms normally invest a very small portion of their R&D budget to fundamental research. Table 6.4 presents some figures on the portion of total public R&D expenditure on science in OECD countries.

From the table it can be derived that in most countries expenditure on science, or the 'advancement of knowledge', is not the largest part of public R&D. Defense is very important in some countries. 'Industrial development' includes direct financial support for R&D, discussed below. Behind such figures country-specific institutions are hidden. We discuss this matter in the chapter on 'national systems of innovation'. Despite these differences the tendency exists for governments to create similar instruments to stimulate and co-ordinate scientific activities. Examples are funding schemes for basic research that are more directed towards performance than in the past, and centres of excellence. Furthermore, countries increasingly concentrate research efforts on areas of particular scientific or technical relevance, especially the so-called generic technologies (OECD, 1994, 22).

Technology policy

From the data in Table 6.4 it became clear that expenditure on science is lower than on industrial development. In most countries the two main areas of industrial development are 'technology policy' and 'regional policy'.

Table 6.4 Expenditure on science and other areas as a percentage of total public R&D in selected OECD countries

	Industry			Government			High education			Private non profit		
	1981	1985	1991	1981	1985	1991	1981	1985	1991	1981	1985	1991
United States	70.3	72.6	68.2	12.1	11.7	11.4	14.5	12.8	16.9	3.1	3.0	3.4
Canada	48.7	53.1	53.8	23.4	21.8	18.8	27.0	24.1	26.4	0.8	1.0	1.1
North America	69.3	71.7	67.5	12.6	12.1	11.8	15.1	13.3	17.4	3.0	2.9	3.3
Japan (adj.)	66.0	71.8	75.4	12.0	9.8	8.1	17.6	14.2	12.1	4.5	4.2	4.4
Belgium [4]		71.5	72.6		5.5	6.1		18.7	17.4		4.3	3.9
Denmark	49.7	55.3	58.5	22.7	19.5	17.7	26.7	24.4	22.6	0.9	0.8	1.2
France	58.9	58.7	61.5	23.6	25.3	22.7	16.4	15.0	15.1	1.1	1.0	0.8
Germany	70.2	73.1	68.9	13.7	12.9	14.9	15.6	13.5	15.8	0.5	0.4	0.4
Greece	22.5		26.1	63.1	53.2	40.1	14.5		33.8			
Ireland	43.6	51.3	62.0	39.3	27.6	13.7	16.0	19.9	22.6	1.1	1.2	1.6
Italy	56.4	56.9	58.5	25.7	23.9	21.5	17.9	19.2	20.1			
Netherlands	53.3	56.2	53.2	20.8	18.3	19.6	23.2	23.2	24.7	2.8	2.3	2.5
Portugal [1,2,4]	28.6	29.6	26.1	47.3	41.3	25.4	19.9	24.6	36.0	4.2	4.5	12.4
Spain	45.5	55.2	56.0	31.6	24.2	21.3	22.9	20.6	22.2			0.5
United Kingdom	63.0	63.3	63.9	20.6	18.0	15.5	13.6	14.5	16.6	2.8	4.3	4.1
EC	62.5	64.0	63.4	19.5	18.7	18.0	16.6	15.7	17.3	1.4	1.5	1.4
Austria [3]	55.8	54.8	58.6	9.0	8.4	7.5	32.8	34.9	32.4	2.3	2.0	1.6
Finland	54.7	58.7	57.0	22.5	19.9	20.2	22.2	20.9	22.1	0.6	0.5	0.7
Iceland	9.6	15.4	21.8	60.7	48.3	44.5	26.0	30.0	29.4	3.7	6.3	4.4
Norway	52.9	62.7	54.6	17.7	14.4	18.8	29.0	22.2	26.7	0.5	0.7	
Sweden	63.7	68.0	68.2	6.1	4.4	4.1	30.0	27.4	27.6	0.3	0.2	0.1
Switzerland [3]	74.2		74.8	5.9		4.3	19.9		19.9			0.9
Turkey			21.1			7.9			71.1			
Nordic Countries	58.5	63.7	62.0	12.9	10.7	12.0	28.2	52.2	25.6	0.5	0.4	
Australia [2,4]	25.0	30.0	40.0	45.1	39.7	32.4	28.5	28.5	26.2	1.3	1.8	1.3
New Zealand			31.7			49.3			19.0			
Total OECD	65.9	68.8	67.1	15.0	13.8	13.1	16.5	14.8	16.9	2.6	2.6	2.8
OECD Median	53.3	56.9	58.5	22.7	19.7	18.8	19.9	20.9	22.6	1.3	1.8	1.5

(1) 1980, (2) 1984, (3) 1989, (4) 1990

Source: OECD (1994, 149)

166

The instruments of technology policy can be divided into direct and indirect instruments. Direct instruments try to stimulate private firms' R&D expenditure directly. Examples are tax facilities and subsidies which reduce the R&D costs of projects for the firm. (Cordes, 1989) Indirect instruments try to create a climate in which the creation, diffusion, and application of technological knowledge are facilitated.

When we look at an overview of the elements of the countries' technology policy in terms of budget it becomes clear that subsidies on R&D and tax facilities are the most important. They are called instruments of technology policy. Table 6.5 gives an overview of tax facilities in a large number of countries.

According to CEC (1992) on average 47 percent of governments' support of the manufacturing industry in the twelve countries of the EU, from 1988–90, was devoted to subsidies and 32 percent to a reduction of taxes. The effects of

Table 6.5 An overview of the main tax policies in several countries affecting R&D

Country	R&D tax credit	Special assistance
Australia		150% funding of R&D
		R&D grants
Belgium		R&D personnel subsidy
Brazil		Exemptions from profit taxes
Canada	20% incremental	
China		
Denmark		
Germany		Investment grants
		Tax credit for R&D equipment
France	50% incremental	R&D grants to selected industries
Hong Kong		
Ireland		
Italy		
Japan	20% incremental	Trade policies beneficial to R&D equipment
Netherlands		Subsidy for R&D labor and capital
Norway		Expense against future R&D
Singapore		200% funding of R&D
South Africa		
South Korea		Expense against future R&D
Spain	15% on R&D	
	30% on R&D equipment	
Sweden	30% incremental	Subsidy on R&D salaries
Switzerland		
Taiwan	20% incremental	
UK		
USA	20% incremental on R&E	
	20% on basic research	

Source: Leyden and Link (1993, 22)

both instruments are similar: a reduction in the cost of R&D for the innovating firm. Tax credits allow for a postponement or reduction of taxes to be paid on R&D activities (income taxes in the case of labor cost of R&D, profit tax in the case of capital expenditure on R&D). Subsidies directly finance a part of the R&D cost. Van den Berg et al. (1990) analyze a Dutch scheme, where a percentage of R&D labor cost is subsidized ex post.[8]

A similar subsidy scheme has been used in Germany from 1979 to 1987. Like the Dutch scheme it was especially aiming at an increase of R&D activity in small and medium-sized firms, which, depending on the definition used, range from 10–500 employees. In the first year the number of applicants was already 4–500 enterprises and by 1983 this number had grown to 7–500 (Meyer-Kramer, 1995, 610–11). Meanwhile the amount of government money involved was more than DM 3.0 billion. The evaluation of this policy instrument estimated the effect of the subsidy scheme as 60 percent. This meant that 60 percent of the projects were undertaken because of the subsidy available, while 40 percent would have been undertaken without subsidy. Most R&D activity stimulated by this German policy scheme can be typified as 'development'. The aim of most development projects was improvement and redesign of existing products (ibid., 614).

Generally two kinds of cost are involved in specific instruments for technology policy: (1) the costs of the subsidies or taxes themselves, that are often reflected in a fixed budget per year; (2) the organizational costs of carrying out the instrument, such as the costs of the office which administrates the preparation, execution, and control of the instrument. In many analyses of technology policy instruments, the organizational costs are not taken into account. Instruments may differ greatly with respect to the complexity and costs of the organization, though. In Table 6.5 organizational costs of the various instruments have not been presented.

6   THE THEORETICAL SIDE OF SCIENCE AND TECHNOLOGY POLICY

In Chapter 5 we have discussed that, related to the appropriability problem, a difference exists between social and private benefits of innovation. Innovation projects for which the private net benefits are smaller than a minimum level ($b_p < b^*$) but the social net benefits are considerably greater ($b_s \gg b_p$) are not carried out by private firms. Nelson (1959) explained why, as a result, fundamental research projects are largely financed by government. In the first part of the chapter patents were supposed to increase the private benefits of innovation projects by reducing their market uncertainty. Another way of stimulating private investment in R&D is by reducing the costs of innovation projects, which implies a mitigation of the effects of technological uncertainty.[9]

The theoretical arguments for publicly financing fundamental research and for government stimulation of applied research and development projects are discussed in the next section.

The public finance of basic research

'Basic research' or 'fundamental research' has been defined in Chapter 1 as the activity directed towards the advancement of knowledge in general. From an economic point of view Nelson (1959) identifies the following characteristics of basic research: uncertain outcomes (compare with Table 4.6), a longer time horizon than private firms normally use, potential commercial applications in areas unrelated to the original field of study. These characteristics make it clear why basic research is to a large extent financed publicly. Nevertheless, a minor part of the R&D budget of private firms is spent on fundamental research. Why and under what conditions?

Let us as an example define a basic research project P_b as a project where the probability that a basic scientific discovery D is done is s (the probability of failure is therefore 1-s). Let us further assume that the knowledge underlying D is completely public and that on its basis N new products are developed by M private firms, each of them operating in its own market M_m. In each market there are K firms which start a project aiming at the development of a new product on the basis of D, but only one succeeds.[10]

Let us now confront this situation with a multidivisional firm that undertakes project P_b. This firm has L units/divisions (L<M) each focused on a specific market M_l. The chance that a unit is the first in its market to develop a new product on the basis of D is 1/K.[11]

It becomes clear from this example that the social benefits of D are substantially greater that the private benefits of the multidivisional firm. Depending on the R&D expenditure (P_b and the other innovation projects) and the probability of success per market ($^s/_K$) the private multidivisional firm may not undertake the basic research project. One reason is that it lacks the market-technological expertise for certain applications (for the M-K markets) or sufficient expertise for success in other applications (in the K markets) which can be seen as a form of market uncertainty. The other reason is that the technological uncertainty of the basic research project itself is too high (s is small). In any respect the social benefits of the project will be greater when it is carried out by a public research institute (s per market compared to $^s/_K$).

So basic research is to a large extent financed by government because private firms find most projects too uncertain and the social benefits of some projects are very high (but we do not know which ones and the benefits may not be seen for a long time). Nevertheless a large number of applied research and development activities are based on, or at least profit from, basic research results in the past.

If private firms are carrying out basic research they are probably either optimistic about the possible success in technological terms (s) or they control a large enough number of markets to appropriate a sufficient part of the total net benefits, that is the ratio of benefits of the K new products and the total M applications of the basic research project (the Nelson argument). Related to a relatively high probability s one could ask the question 'how basic is the basic research carried out by private firms?'. In Chapter 1 we defined generic research as research that was perceived as having immediate applications in commercial markets. We press the argument here further by stating that this kind of research is necessary to even undertake certain innovation projects. The result of a generic research project Pg is then to build a knowledge base with which one can 'enter the field' of a series of new product development projects. Without such generic research there is no chance of successfully completing any one out of a number of new product development projects $\{Pd_1, \ldots Pd_N\}$. So if s is now the probability that a firm does acquire the generic knowledge base which is conditional to undertake any applied research project, then it holds that the probability of success on commercial application in a specific market (q) without such a base is zero . . . $q/1-s(Pg) = 0$. This would imply that private firms are undertaking generic research projects (which we assume to be part of the basic research actually carried out by private business) if they can only enter certain fields of application by successfully completing such projects. Policy and strategy reports regularly mention microelectronics/information technology, biotechnology, and new materials as such generic technologies (see Mansfield et al., 1977, for example, for a discussion of the topic of this section).

Government's role in stimulating private R&D

A large majority of the R&D budgets of private firms is spent on applied research and development that by definition concern projects with the goal to commercially apply technology. The difference with basic research is that commercial application is falling within the used time framework for (R&D) investment projects and within the domain of market expertise of specific firms, although in reality it may be a matter of degree rather than a principal difference. Theoretically one would like to stimulate those projects where $b_p < c_p$ and $b_s > b_p$ (and $b_s > c_p$). We assume here that projects with high technological uncertainty are included, because such projects may lead to high cost estimates.

It is discussed here whether the two main instruments presented in the empirical section, subsidies on R&D and tax facilities, are indeed reducing technological uncertainty, and if so, whether the benefits of this reduction are greater than the costs (that is, the costs of providing the instruments). To that purpose we first discuss some basic properties of these technology instruments.

Generic and specific technology policy instruments

Whether government is able to judge 'theoretical properties' of innovation projects, such as the difference between social and private benefits, depends on the way it handles projects which are submitted for public support, mostly resulting in cost reduction for the private firm. In this respect one can make a distinction between generic and specific technology policy instruments.

A generic instrument is in principle accessible to any firm which undertakes an innovation project, while in the case of a specific instrument public support depends on an evaluation of the project proposal. A specific instrument could, for example, be focused on a specific technology such as biotechnology and only firms with an innovation project in this area can apply for support. The advantage of a specific instrument is that the number of potential applicants is normally small and that a priority list of all submitted projects can be made. The disadvantage is that prior to any project proposal government makes a decision to support firms in one area (for example a specific technology) and to exclude all other firms. Another disadvantage is that for a selection of projects government needs specific information about the contents of each project. One could argue that government officials will not have as much knowledge on the project as the firm itself.

Once government decides that an instrument should be generic it consequently will be confronted with large numbers of application. This limits the possibility to evaluate each project proposal properly. If we assume for simplicity that all innovation projects can be divided into N technology areas with on average M projects, than government needs to evaluate N times more proposals in the case of a generic instrument than in the case of a specific instrument. An additional problem might be that it has to be decided whether a project in technology area T_1 is more valuable for society than a project in area T_2. This will usually be more difficult than a comparison of two projects in the same technological area.

In practice generic instruments require hardly any specific information about the contents of the innovation projects (Meyer-Kramer, 1995; van den Berg et al. 1990). The goal of such instruments is to stimulate R&D in private business in general by offering some financial incentive. The main question is then to what extent projects will be undertaken by private firms for which holds that $b_p < c_p$ and $b_s > b_p$ and not projects with $b_p > c_p$ and $b_s = b_p$. In contrast, if in case of a specific instrument government is evaluating different project proposals, how can it acquire information about private and social benefits and costs of each project?

A cost–benefit analysis of tax and subsidy instruments

In the beginning of the chapter we argued that the administration costs of the instrument must be included in the cost estimations. Apart from this complication estimating the costs of a tax credit or a subsidy scheme is straightforward: the total cost is the sum of the subsidies given to each of the firms (or the total amount of tax not received, because of the tax allowance).

The complications clearly arise when estimating the benefits of the instrument. Two main obstacles can be mentioned: the measurement of benefits of an innovation project in general and the effect of the instrument on these benefits. We discussed the difficulties for the firm to make proper estimations. When such estimations are available, government is dependent on the firms to provide them with the data. An analysis of the efficiency of technology policy instruments is therefore dependent on the firms' cooperation.

As with the calculation of a firm's innovativeness the efficiency of an instrument can be estimated along two lines: the input of innovation and the output of innovation. With the input approach one looks at the effect of the subsidy on the R&D expenditure of the firms participating in the scheme. The two subsidy schemes mentioned in the empirical section have been evaluated in terms of an increase in inputs (Meyer-Kramer, 1995; van den Berg et al., 1990). The output approach focuses on the technological or commercial results of innovation projects. With regard to subsidy and tax schemes output evaluations are seldom available.

We take a fictitious example of a subsidy scheme on innovation to discuss the efficiency of technology policy instruments. This makes clear a number of problems to evaluate in practice.

Assume that in a particular country or region the government introduces a subsidy scheme for innovation in 1998. Ten thousand firms exist in the economy, of which 1,000 undertook at least one innovation project in the year previous to the subsidy scheme. The average cost of an innovation project is 100,000 units of the national currency. Let us assume that total R&D expenditure by private firms was 150 million in 1997 (an average of 1.5 project per firm). In 1998 500 firms receive a subsidy of 25 percent on the cost of one of their R&D projects. The total amount of subsidy per project is 25,000, so government is spending 12.5 million. The instrument can be said to be extremely efficient when all subsidies go to projects that have been carried out in addition to the projects to be expected on the base of R&D expenditure in the previous year. This is the case when government observes that private R&D expenditure in 1998 rises from 150 million to 187.5 million (excluding the 12.5 million of government subsidy); the 500 firms receiving a subsidy together spend 37.5 million extra on R&D.

The government could accept a lower efficiency rate, but it is very difficult to give theoretical arguments what specific 'lower level' would still be acceptable. It is clear that a rise of R&D expenditure with 12.5 million in 1998 is inefficient, because this would mean that no extra innovation projects have been initiated due to the subsidy scheme. The judgment whether an instrument is efficient or not of course also depends on what comes out of the subsidized projects. If, for example, the number of patents rises with the same rate as R&D expenditure in 1998, it may be implied that the quality of the subsidized projects equals the average quality of the projects carried out on the base of private firms' own cost–benefit analyses.

What can be concluded when private R&D expenditure in the whole economy rises to 237.5 million instead of 187.5 million? Strangely enough this could imply that the instrument is completely inefficient. This is the case when R&D expenditure of the firms without a subsidy rises from 75 million in 1997 (50 percent of total R&D) to 125 million in 1998. This situation can be interpreted as follows. Apparently many firms in the market are confronted with an increased need to innovate, translated into an average rise of R&D expenditure in the firm of 66 percent. Fifty percent of the innovating firms take advantage of the subsidy scheme, by receiving a 25 percent subsidy for an innovation project that they would have undertaken anyway.

Up to here the example assumes that no 'death' or 'birth' occurs within the population of firms and that the number of innovating firms as a percentage of the total number of firms is not changing either. Let us assume now that the same amount of subsidy (12.5 million) goes to existing innovating firms as well as newcomers. Incumbents receive 6.25 million and another 6.25 million goes to firms that start their first R&D project. Let us assume, for simplicity, that the size of the projects of the newcomers is the same as for the old firms. The efficiency of the subsidies going to the newly innovating firms is very high: for each million on subsidy private expenditure rises with 3.0 million. From an output point of view this situation could be less satisfactory.[12] Firms without an experience in innovation can be expected to be less successful, on average. The aim of many tax or subsidy instruments on innovation is to stimulate R&D activity of small firms. If this was the case in our example, the instrument can be assessed as very effective, because 50 percent of all the subsidized projects are undertaken by firms without any previous innovation record. One should also look at the long run, however. As we argued at the beginning of the book, successful innovations often require years of accumulated R&D expenditure. Most studies on R&D subsidy or tax allowance schemes do not cover a long period in which R&D expenditure and innovation output indicators for both subsidy receivers and other firms can be compared. This makes a cost–benefit analysis of technology policy hazardous.

Generally it can be said that taxes and subsidies stimulate R&D expenditure, not the undertaking of successful projects. It could even be argued that small and medium-sized firms, the majority of firms in the economy, need assistance in the selection and management of innovation projects as much as financial support. This is part of the 'modern view' on technology policy, that is discussed in the chapter on national systems of innovation.

Conclusions with respect to technology policy

In practice many industrialized and industrializing countries have tax or subsidy schemes to stimulate R&D activity. Tax allowances are by definition generic in the sense that every firm with R&D expenditure can receive a favorable fiscal treatment. Subsidy schemes can be generic or specific. Specific instruments often cover the technology fields that have been called generic in Chapter 1: biotechnology, microelectronics, and the like. Innovations stimulated in these fields are expected to have large 'spillovers' to many markets in which these innovations are used or further developed. In generic schemes government is not evaluating the value of each subsidized project. In specific schemes government usually does make an assessment, because projects are granted a subsidy or not. In addition projects are sometimes prioritized and one project receives more subsidy than another.

In theory stimulating those projects which have a net social benefit but a net private loss is the most effective way of implementing technology policy programs. In general it is difficult for firms to make refined cost–benefit analysis of their innovation projects. In addition each selected project needs to fit the overall strategy, as we have seen in Chapter 3. Boundedly rational behavior of innovating firms also implies that projects that are undertaken as a result of some technology policy instrument need not be economically viable. Apart from this rationality issue an assessment of technology policy instruments turned out to be very difficult. Our example showed how different assumptions lead to quite different outcomes in terms of (input) efficiency and effectiveness.

NOTES

1. I am indebted to Peter van As who wrote his master thesis 'Differentiatie in de Patentduur' (Differentiation in Patent life) under my supervision. The high quality of his work provided me with deeper insight and an extensive list of references on the issue of patents in economics.
2. As we stated earlier R&D may not be the only activity of the firm which results in innovations.
3. In addition to the costs to apply for a patent.
4. The first step was making the procedures to get a patent in a certain country, once the inventor received a patent in another country, simpler.
5. Of course, this behaviour may be influenced by other factors, such as the size and the age of the firm, and so on.
6. This diversity could take several forms. For example competition in product design, which increases the chance that the market produces a better new product than in the case of homo-

geneous firms. Another possibility is that one or more firms create an innovation which could (also) be commercially applied in another industry (see the explanation of Pavitt's technological taxonomy of industries in Chapter 3).

7. In the long run the first mentioned issue may be more serious, namely, a loss of competitive managers in the country.

8. This implies that the firm has to finance the innovation project itself, and only afterwards, in this case each half year, does the government subsidize a maximum of 40 percent of the labor costs over that period. For more facts about subsidies and tax facilities see Mansfield (1986), Mansfield and Switzer (1985), Mansfield et al. (1977, 1981), and Meyer-Kramer and Montigny (1990).

9. As explained before, technological uncertainty concerns 'cognitive' problems one may encounter during an innovation project. These unknown problems imply that at best estimations of the costs and speed of an innovation project can be obtained prior to completion of the project. For simplicity we will speak of a reduction of technological uncertainty when the costs of R&D are reduced by the use of government instruments.

10. The winner, for example, gets a patent or is otherwise strongly protected.

11. We assume here that all units are stimulated to undertake new product development projects on the basis of D.

12. Note that in the case of this output growth of innovation, as well as in the formerly discussed case of growth of R&D expenditure in 1998, it is assumed that the 'autonomous' growth rate of private innovative activities is zero or constant at a positive level.

REFERENCES

Basberg, B.L. (1987), Patents and the measurement of technological change: a survey of the literature, *Research Policy* 16, 131–41

Berg, M.C. van den, A. van Dijk and N. van Hulst (1990), Evaluating a Dutch scheme for encouraging research and development, *Small Business Economics* 2, 3, 199–211

Commission of the European Community (1992), *Third Overview of Support Measures of Member Countries for Manufacturing Industry and Specific other Sectors in the European Community*, Luxemburg

Cordes, J.J. (1989), Tax incentives and R&D spending: a review of the evidence, *Research Policy* 18, 119–33

Gilbert, R. and C. Shapiro (1990), Optimal patent length and breadth, *Rand Journal of Economics* 21, 106–12

Jewkes, J., D. Sawer and R. Stillerman (1969), *The Sources of Invention*, London, Macmillan

Kaufer, E. (1989), *The Economics of the Patent System*, Chur, Harwood Academic Publishers

Levin, Richard C., Alvin K. Klevorick, Richard R. Nelson and Sidney D. Winter (1987), Appropriating the returns form industrial research and development, *Brookings Papers on Economic Activity* 3, 783–820

Lerner, J. (1994), The importance of patent scope: an empirical analysis, *Rand Journal of Economics* 25, 2, 319–33

Leyden, D.P. and A.N. Link (1993), Tax policies affecting R&D: an international comparison, *Technovation* 13, 1, 17–25

Mansfield, E. (1984), R&D and innovation: some empirical findings, in Z. Griliches (ed.), *R&D, Patents, and Productivity*, Chicago, Chicago University Press

Mansfield, E. and L. Switzer (1985), The effects of R&D tax credits and allowances in Canada, *Research Policy* 14, 97–107

Mansfield, E., J. Rapoport, S. Wagner and G. Beardsley (1977), Social and private rates of return from industrial innovations, *Quarterly Journal of Economics* 16, 221–40

Mansfield, E., M. Schwartz and S. Wagner (1981), Imitation costs and patents: an empirical study, *Economic Journal* 91, 907–18

Merges, R.P. and R.R. Nelson (1994), On limiting or encouraging rivalry in technical progress: the effect of patent scope decisions, *Journal of Economic Behavior and Organization* 25, 1–24

Meyer-Kramer, F. (1995), Technology policy evaluation in Germany, *International Journal of Technology Management, Special Issue on the Evaluation of Research and Innovation*, 10, 4/5/6, 601–21

Meyer-Kramer, F. and P. Montigny (1990), Evaluations of innovation programmes in selected European countries, *Research Policy* 18, 313–32

Napolitano, G. (1991), Industrial research and sources of innovation: a cross-industry analysis of italian manufacturing firms, *Research Policy* 20, 171–78

Nelson, Richard R. (1959), The simple economics of basic scientific research, *Journal of Political Economy* 67, 3, 297–306

Nelson, R.R. and Winter, S.D. (1982), *An Evolutionary Theory of Economic Change*, New York, Belknap Press

Nordhaus, W.D. (1969), *Invention, Growth, and Welfare: A Theoretical Treatment of Technological Change*, Cambridge, MIT Press

Nordhaus, W.D. (1972), The optimum life of a patent: reply, *American Economic Review* 62, 428–31

OECD (1994), *Science and Technology Policy; Review and Outlook 1994*, Paris, OECD

Pavitt, K. (1982), R&D, patenting and innovative activities, *Research Policy* 11, 33–51

Rosenberg, N. (1976), *Perspectives on Technology*, Cambridge (MA), Cambridge University Press

Schankerman, M. and A. Pakes (1986), Estimates of the value of patent rights in European countries during the post-1950 period, *The Economic Journal* 96, 1052–76

Scherer, F.M. (1983), The propensity to patent, *International Journal of Industrial Organization* 1, 107–28

Scotchmer, S. (1991), Standing on the shoulder of giants: cumulative research and the patent law, *Journal of Economic Perspectives* 5, 1, 29–41

Schmookler, J. (1966), *Invention and Economic Growth*, Cambridge (MA), Harvard University Press

Taylor, C.T. and Z.A. Silberston (1973), *The Economic Impact of the Patent System: A Study of the British Experience*, Cambridge, Cambridge University Press,

Teece, D.J. (1986), Profiting from technological innovation, *Research Policy* 15, 6, 78–98

7. The firm as a learning organization

1 INTRODUCTION

In neoclassical theory, where the focus is predominantly on the market, the firm is seen as a production function in which inputs are transformed into outputs with the help of a given technology. 'Learning-by-using' plays a role in the sense that unit production costs are assumed to diminish when the number of products made rises. Learning how to 'create new production functions' is not considered. Transaction cost theory (Williamson, 1985) introduces an organizational element in the theory of the market by assuming that the firm is a governance structure that efficiently administers internal transactions (that remain hidden in the production function). In this theory it is stressed that decision makers have limited abilities. As a result contracts with other parties do not contain all possible contingencies. This implies that a governance structure must be designed to allow for mutual adjustments by the transacting parties, when required. We come back to this issue in Chapter 9. Here it suffices to state that some organizational learning is assumed, although implicitly. Evolutionary economists (Nelson and Winter, 1982) perceive the firm as a 'bundle' of routines that enable the replication of existing products and production processes. With regard to innovation they make a distinction between a regime of cumulative technology and a regime of discrete technology. In this economic-evolutionary theory learning with respect to innovation plays a central role. In regimes of cumulative technology innovation is the result of the accumulation of knowledge, following from years of R&D in a specific field of inquiry. Innovation is to a large extent based on routines too. In regimes of discrete technology markets are characterized by radical innovation and firms are confronted with high uncertainty and the need to learn radically new knowledge. Innovation is then best characterized as a deliberate search process for new technology, where trial and error are quite dominant. In both regimes the firm represents a 'knowledge creating entity'. In the one case it is knowledge that has been gradually built up, in the other it is radically new knowledge. The organizational and individual requirements may be completely different in each situation.

In the first part of the book it has become apparant that creating innovations is not a trivial activity that merely depends on the financial ability to carry out R&D and on the existence of the proper 'appropriability conditions'. In the discussion of patents, for example, it turned out that imitation of new technology and the diffusion (transfer) of knowledge in general can be rather problematic for private as well as public organizations. The recognition of barriers to knowledge transfer, that is, the recognition of a world without a free flow of information and, more importantly, a world where 'flowing' information is difficult to comprehend, is one of the basic arguments for a chapter on the economic aspects of learning with respect to innovation. The 'economics of learning' is therefore about the perception, selection, and commercial use of technological information. How private firms learn, and how they organize the processes of creation and implementation of innovation are the main topics. An important aspect of this technological learning process is the use by the innovating firm of outside information sources. While the original economic contributions tended to see the firm as an isolated organization, today innovators are seen as an actor in a web or network of institutions which may facilitate or hamper the innovation process.

The problem

In earlier chapters innovation by a firm was seen as the successful completion of one or a number of innovation projects. Illustrative of the traditional economic approach was the perception of the innovation process as the allocation of a firm's R&D resources across a number of potential projects with given estimated costs and benefits. For one thing, projects may fail and the costs of such projects must be incurred in other, successful projects. A more important point for the current topic of learning is the question whether the firm has the know-how to undertake all the projects with high expected net benefits. And, more fundamentally, how does a firm get the information, or better, the insight or know-how to formulate potential projects and to estimate costs and benefits? Or in other words: 'how does the firm learn which projects should be undertaken?'. One can imagine that engineers or researchers in a firm tend to come up with new project proposals that fit the knowledge and experience build up in several previous innovation projects. If at all able to formulate radically new projects, how can the firm undertake projects for which totally different know-how is required? Is the labour market sufficiently transparant to hire people with the needed skills? We have already argued that it usually is not. The distinction between radical innovation and incremental innovation is nevertheless a matter of degree rather than 'principle'. Incremental innovations also require some new knowledge that has to be learned and the same 'learning problems' occur, be it in a slighter degree. One of the main problems in the whole process of innovation is that new knowledge is the input and at the same time the output.[1] We quoted Mowery and Rosenberg to make clear that in their opinion economists concentrated too much on the right market and appropriability conditions for innovation, and neglected the innovation process, which in itself can be quite problematic.

Technological learning is a subset of learning in general (Dodgson, 1993; 1996). The results of any learning process depend on the abilities of the 'students', the contents of what must be learned, and the way in which the information is structured and explained during the learning process (related to the ability of the teacher, but there is no teacher in most innovation processes). Instead of information being explained the innovation process can be characterized by a search process of the researchers and engineers. They make use of internal and external information sources, such as university laboratories, patent information, ideas from employees. R&D can be seen as the main vehicle to translate information from various sources into new products and processes. Successful innovation depends as much on interfaces with marketing, production and other functional activities in the firm as on the R&D process itself. Technological learning is therefore also dependent on the firm's ability to learn in other areas and to coordinate the things learned in various areas. We limit this chapter largely to technological learning and R&D per se. Figure 7.1 reflects our findings up to this point.

Figure 7.1 The basic flows of information

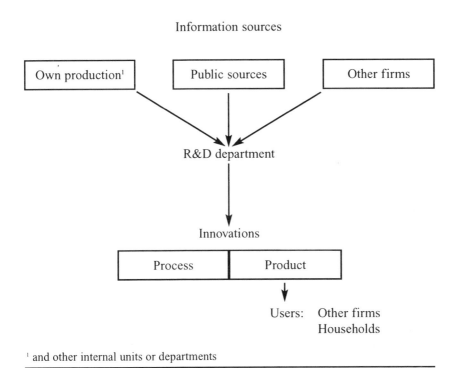

¹ and other internal units or departments

A final point in the introduction to the learning problem in innovation is the distinction between 'learning the right things' and 'learning the things right'. Learning the right things has to do with strategic choices of the firm on the kinds of technology one plans to learn. Learning the things right is about procedures. The routines of evolutionary economics are about procedures to learn. The rapidly growing body of management literature on learning is also stressing 'how to organize the firm' in such a way that continuous learning by individuals (and the organization as a whole) is promoted. Nevertheless learning must always be about both aspects.

In the empirical section we examine how much we know about this kind of technological learning. We look at the accumulation of know-how that follows from undertaking years of R&D, at the use of internal sources of innovation, but especially at the use of external sources of innovation. The theoretical section explores what kind of concepts are currently used to handle learning in innovation.

The chapter intends to provide an answer to two questions that have been raised in the earlier chapters of the book: (1) What are the main problems today for innovating firms to continuously create and implement innovations? (2) What role does the organization, and especially organizational change play in this? These two questions are covering basic topics that, as we argued, need to be seen as complementary to the theoretical discussions by neoclassical economics, transaction cost economics, and, to a lesser extent, evolutionary economics of appropriability of innovation profits.

2   THE EMPIRICAL SIDE

How do firms learn new technological things? Basically in two ways. First, they undertake systematic research into new technological opportunities and some of these opportunities are developed further into new products and processes. Second, firms go by trial and error, people are confronted with specific problems in their production process, with ideas from customers with regard to product ideas or practical ideas from a variety of employees. How firms pick up ideas and how these are channelled to the R&D department and even the more systematic execution of research and development activities, are to a large extent based on tacit knowledge (Senker and Faulkner, 1996). The knowledge embodied in R&D activities can further be described as specific, complex, and cumulative in its development (Pavitt, 1987). This implies that a 'typical' innovation project can only be carried out when the team members have several years of experience in the relevant technological field(s) and ideas and problem-solving methods used in the project will, on many occasions, be linked to previous projects of the firm. This does not exclude the possibility that a large part of the technological information in an industry is common to all firms (Gibbons and Johnstone, 1974). It is the ability to learn specific things 'translated' into innovations from this common pool of knowledge that may distinguish the average from the innovative firm in the industry.

Here a distinction must be made between learning new things that fit with the current expertise and experience of the firm, and learning completely new things. One way to discover how completely new things are learned is to compare a firm with an established innovation record in a specific technological field with a firm that has to start its first innovation project in that field. In the discussion of Henderson and Clarke (1990) it became clear that specialized procedures and organizational structures of a (successfully) innovating firm could hamper the effective learning of something completely new. Organizational aspects are thus important in understanding technological learning.

Sources of information

Today the use of sources of information is seen as a major contributor to innovation and learning. Table 7.1 provides data on the use of internal and external sources by innovating firms.

From several studies on the use of external sources of information (Napolitano, 1991; Brouwer and Kleinknecht 1994) it follows that part of the innovating firms do not undertake R&D and rely completely on the use of external sources. This raises the question how and by whom the information from these external sources is 'transformed' into innovations. Normally it is assumed that, although various departments of the firm may be involved in the acquisition of external information, the R&D department is ultimately responsible for the realization of technological innovations on the base of this information. Apparently, other 'functions' in the firm can produce innovations too. In our discussion of different team structures in Chapter 3 we argued that sometimes only people from engineering, production, and marketing are involved in product development. It must be admitted, however, that not enough empirical studies are available to completely explain this absence of R&D in innovating firms. Two questions can be raised: Which kinds of functions/activities except internal R&D are responsible for innovations? What do we know about the kind of innovations that are produced without R&D?

From Table 7.2, which is based on the data of Napolitano (1991), several conclusions can be derived. First of all, only 31 percent of the 8,220 Italian firms in the survey, that reported innovations undertake R&D. Napolitano (ibid., 172) states that this means formally registered R&D. Second, in most industries R&D expenditure is a minor portion of total innovation costs. Further analysis of the data shows that design and engineering on average account for 25 percent and production investment for 51 percent of total innovation costs.[2]

Table 7.1 Percentages of firms in the Netherlands with innovative activities in 1990–92 which acquired technological information in 1992

	General	License	Public	Outsourcing	other	Consultant	JV	Acquisition of firm	Purchased equipment	Informal contacts	Hiring qualified personnel
						From Source:					
Size class											
Manufacturing											
10–19	38	1.2	8.9	6.8	14.5	0.0	0.0	2.1	13.8	12.8	2.4
20–49	39.8	4.9	9.3	4.8	6.4	13.2	1.6	2.9	17.4	17.4	7.9
50–99	45.4	7.8	14.1	7.1	10.8	10.7	3.0	1.4	15.5	19.3	10.2
100–199	59.3	11.2	20.4	13.3	15.8	16.5	3.8	2.9	22.1	22.9	14.7
200–499	65.4	8.9	26.5	17.5	26.2	18.8	5.3	6.2	22.6	26.6	14.7
>500	72.9	17.1	40.3	14.4	28.6	24.9	16.9	9.3	25.9	26.6	22.9
Total	**45.9**	**6.1**	**13.8**	**8.0**	**12.3**	**13.2**	**2.6**	**3.0**	**17.6**	**18.5**	**9.0**
Services											
10–19	33.6	2.9	4.5	2.8	4.2	8.3	0.1	0.5	11.2	13.9	8.3
20–49	40.4	7.4	9.7	5.7	4.3	13.8	3.4	2.2	13.3	16.4	14.8
50–99	37.3	7.2	7.9	4.8	8.6	13.4	2.7	2.2	11.8	15.7	11.5
100–199	54.8	10.0	15.1	10.2	14.7	18.2	5.4	1.3	17.3	23.1	14.8
200–499	59.3	8.1	21.7	13.3	17.8	21.6	6.7	6.8	20.4	23.4	18.7
>500	69.3	15.0	28.4	11.6	22.2	33.8	14.9	9.9	20.9	29.6	25.1
Total	**38.5**	**6.8**	**7.2**	**4.4**	**4.5**	**13.7**	**3.2**	**1.6**	**11.2**	**16.1**	**14.8**
Overall	**41.1**	**6.6**	**9.5**	**5.7**	**7.3**	**13.5**	**3.0**	**2.1**	**13.4**	**16.9**	**12.8**

Source: Brouwer and Kleinknecht (1994)

Industries with the highest percentage of R&D in total innovation costs are those which are normally considered as the most innovative (chemicals, petrochemicals, synthetic fibers, office machinery computing, electric/electronic industry, precision instruments all have scores >50 percent).

This can be interpreted as if in the most innovative industries a relatively large part of the innovations are radical innovations, for which R&D is required, although other sources may be used too. In the less innovative industries the large majority of innovations are incremental ones, that often take place without any formal R&D. It must be admitted, however, that no data are presented

Table 7.2 Innovating firms, R&D expenditure and innovation costs in Italian manufacturing industry

Sector	Number of innovating firms 1981–85	Percentage of firms performing R&D in 1985	Percentage of R&D in total innovation costs
Petrochemicals	19	52.6	13.8
Metals	148	31.8	6.6
Non-metals, minerals	583	30.4	5.0
Chemicals	490	70.8	30.8
Synthetic fibres	11	54.5	12.2
Metal products	1051	20.0	7.3
Mech. engineering	1404	44.7	16.0
Office machinery computing	11	90.9	34.7
Electric, electronic	658	58.7	24.3
Motor vehicles and parts	177	49.2	15.5
Other transport	100	47.0	4.1
Precision instruments	144	52.8	23.7
Food	319	17.6	3.6
Sugar, drinks	212	24.1	10.3
Textiles	714	13.6	6.4
Leather	132	15.9	4.6
Footwear, clothing	412	9.5	5.8
Wood, furniture	575	13.0	7.6
Paper, printing	434	12.9	4.3
Rubber, plastic	477	33.8	32.0
Other manufacturing	149	19.5	12.4
Total	8220	31.1	17.9

Source: based on table 1 of Napolitano (1991, 172)

on the kinds of innovation in the Italian firms. For the 5,663 firms without any formal R&D 'purchase of equipment' is clearly the most important source of innovation.[3] This could mean two things. First, engineers of the purchasing firm improve equipment from the supplier (this would mean a process innovation). Second, new parts bought from a supplier are built into the product, and again engineering staff do the job. If this incorporation of a new part form a supplier does not require any innovative act, the product ought not be registered as a (product) innovation. The study tried to exclude such minor improvements.[4]

One thing is not made clear in this interpretation of the data. Many smaller firms are suppliers to other firms. In most cases of customer–supplier relations the (large) customer develops the component and gives the technical specifications of this new product to the supplier, who merely makes the product. To the extent that the supplier suggests minor improvements about the design of the product to the customer, the supplier may be of the opinion that he innovates. Such customer 'initiated' innovations will probably also be registered as innovations by the supplier in surveys like the Italian one, discussed above. Generally, firms source out specific services and products that are needed in the production process. Every supplier of these products and services is a potential source of information for the innovation process of the main contractor, but of course also the other way around. Monteverde and Teece (1982) discuss some possible disadvantages of information exchange of a customer with its suppliers (this issue is discussed in Chapter 9).

With the earlier mentioned reservations in mind, the study sheds some light on organizational aspects of innovation; some types of innovation may be realized without any R&D, some organizations do not register innovative activities as R&D, although they might actually be R&D activities. What is known from several other studies on the use of external information sources, especially with regard to small and medium-sized firms (SMFs), is that the more innovative a firm is, the more varied and the more 'distant' are the information sources used for innovation. Most SMFs do not undertake innovations and those who do are mainly occupied with minor innovations. These latter firms attach most value to the various 'close' information sources, such as accountants, suppliers, and customers, with which they have regular contact outside the scope of innovation (see Docter and Stokman, 1987).

A specific aspect of the use of external information sources by innovating firms is the information a firm has about the innovation projects and innovation results of competitors. In Chapter 5 on technological competition some figures on the time it takes firms to learn about competitors' research projects were presented. In the management literature gatekeepers, liaison officers and other roles are distinguished, which relate to the acquisition, selection and use of external information.

Radical innovators and new firms

According to the study of Jewkes et al. (1969) the major part of inventions, that ultimately became important innovations, were created by individuals and small firms. From a learning perspective this may be interpreted as follows. The innovation process can be divided into a relatively inexpensive, but 'unstructured' stage of creation and problem solving. This is followed by a stage in which the prototype is further developed and tested. This stage of testing and pilot production is more expensive and can be well structured. Small firms are normally better equipped to carry out the unstructered tasks, while large, formally organized firms have an advantage in the well-structured tasks. Instead of contrasting small and large firms in such a way, it is more important to know how any firm can cope with the contrasting activities 'creativity' and 'exploiting new ideas commercially'.

The organization of learning

In order to promote creativity and (personal) learning some firms stimulate their R&D employees to undertake private projects during working hours by allowing them to to spend a fraction of their formal time on such projects. More generally, it is well known from the literature on management of innovation that creativity is stimulated by allowing 'loosely' organized teams to operate inside the firm. In fact, the 'independent' central laboratories discussed previously in the book are a manifestation of this view. As we have seen, this centralization of research has been followed by a more decentralized organization of innovation activity in large firms. By allowing divisions to contract services from centrally operating R&D departments, creativity is better focused to the commercial needs of the firm. We have also seen that multi-disciplinary product development teams are another organizational adjustment to comply creativity and exploitation. Exploitation of new ideas has to do with learning to use innovations.

Learning to use innovations

In the economics of innovation not only the organizational and managerial aspects of the creation of new products and processes are neglected, but also of their implementation. Technological learning is as much about the effective use of new technological knowledge as about the creation of this knowledge. One must be careful not to draw the wrong conclusions in this matter. The discussion of production functions in Chapter 2 made clear that economists had a clear eye for the commercial exploitation of new technology. In fact, they disregarded the creation process of new technology and focused entirely on the direction of creative effort. By examining price tendencies of the various production factors, firms can anticipate the kind of process technology required to keep manufacturing costs as low as possible. As has been the case in the US and Europe in the 1960s, many firms focused on process technologies that saved on the use of relatively expensive labor. The main limitation of this core approach of economists to innovation is that it cannot be applied to product innovations, that take up the majority of R&D activities. Product innovations are especially dominant in the early stages of an industry's lifecycle. In these stages a relatively large number of relatively small firms compete. In later stages, once a dominant product design defines the 'competitive arena' of the market, large specialized firms may establish the most efficient learning organizations.

In Henderson and Clarke (1990) it was shown that market leaders specialize in all kinds of subtasks to improve the dominant design further. Separate departments focus on a specific technological field relevant to the core product and its components. Each technological field requires specific problem solving procedures and the communication with other technology groups is gradually optimized in terms of interactions required according to the logic of the product design. It was also revealed that this specialized market leader had difficulty in following firms that introduce radical or architectural innovations in the market. In the theoretical section a further explanation of these issues is examined.

3   THE THEORETICAL SIDE

R&D has long been seen as the activity to produce innovations. Between the time when a firm starts a new innovation project and its successful completion the researchers must have learned something new. There are roughly two possibilities: (1) they learned from previous projects or from other knowledge 'in the heads of the researchers' that has not been exploited yet; (2) they learned from outsiders or from knowledge that came from outside. Although it seems obvious to relate innovation to creativity and learning, the economics of innovation has neglected this topic for a long time, and to a large extent still does. The question is therefore legitimate as to what is economical about creativity and learning? As Wheelwright and Clarke left invention outside the organization and planning of new products and processes, and to the extent that those radical development projects demand more in terms of creativity and learning, one could argue that, when these issues are outside the scope of management of innovation how could economics discover patterns in them? Another argument is that basic research is only a minor part of the total expenditure on R&D and that creativity and learning mainly take place there. Creativity and learning can therefore be said to either fall outside the domain of economics or be of minor importance from a cost efficiency point of view. This is, however, too easy. Three remarks make this clear. First, Wheelwright and Clarke kept invention outside the management of new development projects, not outside the management of innovation more generally. They argue that when speed, quality, and low cost are the three crucial features of commercial application of new technology, one must be certain that no elements are incorporated that could easily change estimations about them. Invention clearly could be such a disturbing factor, because as Wheelwright and Clarke (1990) use the term, we talk about radical product and process innovations that have the highest scores on the 'uncertainty list' (see Freeman's table on the degree of uncertainty of innovation types in Chapter 4). Second, the results of new development projects in terms of cost, quality, and speed will be influenced by the kind of 'proven new technology' developed by a separate unit in the firm. From an economic point of view it is therefore interesting to analyze the processes of how to get proven new technology. Third, creativity and learning take place in various innovation activities and not only in those activities related to prototype development. We saw in Chapter 6 that patents are obtained for other results of innovation processes than prototypes. So, clearly, other things than prototypes incorporate new things too, and these new things have had to be learned.

Learning how and learning what

In Chapter 3 the modern view on how to manage innovation has been given. This can be extended to a systematic analysis of how the firm can 'learn how to learn' best from its own innovation experiences, with the goal to continuously come up with innovations and to improve the performance of project teams in terms of cost, quality, and speed. Wheelwright and Clarke (1992, chapters 11 and 12) discuss management tools to improve the firm's learning from its own R&D. It must be stressed that the discussion concern changing existing procedures and creating new procedures to innovate continuously. So we deal here with learning how.

As the economics of innovation has put much more emphasis on causes and consequences of innovation than on the process of innovation itself, it is not surprising that less attention has been paid to learning how. The production function is exemplary of the dominance of learning what. However, as we argued, this emphasis on learning what has a special interpretation: it is concerned with the what in terms of economic consequences, not with the what of technological content. So, firms are assumed to understand that process technologies must be directed at relative prices of production factors, but how one decides what kind of new process technology does the job best is left outside the analysis. Recently some economists developed concepts related to the technological content of innovations. We discuss these in the following sections.

The allocation of resources and resource creation

We have indicated that transaction cost economics (TCE) is an economic theory concerned with the explanation of various forms of organization. One of the main disadvantages of TCE in the perspective of innovation is its focus on the allocation of existing resources: each product or service is matched with a specific governance structure that minimizes costs (especially transaction costs). Strategic plans and actions of the firm are reduced to finding the right organizational solution for existing activities, including R&D. In reality this aspect is only part of top management's concern and, depending on the industry and its life cycle, creating new activities for competitive advantage is more important in the long run. Management literature in the last decades addresses this issue in terms of 'core competencies', 'capabilities', and the incorporation of technology and innovation strategy in the overall strategy of the firm. In economics strategic considerations have long been 'reduced' to the assumption of profit maximization by the firm as a single decision maker. Game theory opened up a large domain for strategic considerations, including R&D activity. This theoretical approach has been concerned with how firms react to each other given a set of activities (with related payoffs). The process of resource creation is not discussed. Quite recently a number of economists have taken resource creation as the 'raison d'être'' of the firm (Kogut and Zander, 1992), although we already suggested that Schumpeter and evolutionary economics took resource creation as a core issue too. This topic of resource creation is further based on Chesnais (1996) who gives a very clear account of the elements of resource creation in economic theory.

R&D as a way to learn

In the empirical section of this chapter we briefly discussed firms that create innovations, but do not undertake R&D. We raised the question there to what extent R&D is needed as a basis to learn from other organizations. We posed the hypothesis that the more complex the innovation aimed for, the more in-house R&D is needed and at the same time the more one is able to learn from specific information sources, usually outside the firm. This is what Cohen and Levinthal (1989, 1990) call 'the double function of R&D'. The first function is the creation of innovations. The second is to build a knowledge base with which one can learn new things from information from, and interpersonal communication with, outside sources. The point is that, while we have been talking about R&D as the unit where innovations are produced, the knowledge thus created at the same time allows research people to perceive technological information 'in' the environment in a specific way. Moreover, R&D may at some time be used mainly to learn from the environment. A firm that undertakes a generic technology project which takes three years does not have innovations in the eyes of the outside world. What it does with its R&D expenditure is building a knowledge base from which, through subsequent R&D expenditure, innovations may result. This issue has also been discussed in relation to the effectiveness of R&D subsidies. There it was argued that subsidies to small firms could be effective, even when the resulting R&D activity does not lead to innovation, because a knowledge base for subsequent innovation projects is laid.

The ability to learn from outside sources is called 'absorptive capacity' and is defined more precisely as (Cohen and Levinthal, 1990, 128) 'the ability of a firm to recognize the value of new, external information, assimilate it, and apply it to commercial ends is critical to its innovative capabilities'. This absorptive capacity is to a large extent a function of the level of prior related knowledge. While Henderson and Clarke (1990) stressed organizational procedures as the cause of learning mainly in the direction of what the organization already knows, Cohen and Levinthal (1990, 129) base their argument largely on cognitive aspects of individuals:

> Research on memory development suggests that prior accumulated knowledge increases both the ability to put new knowledge into memory, what we would refer to as the acquisition of knowledge, and the ability to recall and use it . . . Some psychologists suggest that prior knowledge enhances learning because memory – or the storage of knowledge – is developed by associative learning in which events are recorded into memory by establishing linkages with pre-existing concepts.

This crucial role of prior knowledge not only refers to 'knowing what', but also to 'knowing how'. Prior learning experience enhances the learning of a set of new but related learning skills (ibid., 130):

> Ellis (1965: 4) suggested that students who have thoroughly mastered the principles of algebra find it easier to grasp advanced work in mathematics such as calculus.

Further illustration is provided by Anderson, Farrell, and Sauers (1984), who compared students learning LISP as a first programming language with students learning LISP after having learned PASCAL. The PASCAL students learned LISP much more effectively, in part because they better appreciated the semantics of various programming concepts.

Furthermore, the literature on cognitive structures states that problem-solving skills develop similarly learning skills, despite the fact that learning capabilities involve the development of the capacity to assimilate existing knowledge, while problem-solving skills represent a capacity to create new knowledge. Hence, the ability to solve problems and to create are also dependent on prior

experience with problem-solving and creativity. From the discussion of architectural innovations it became clear that within the firm problem-solving
methods that are specific to the architecture of the product are gradually developed. Once new problem-solving methods are required, the firm is restrained
in its innovative activities.

Learning plays an important role in evolutionary theory. The following
quote from Cohen and Levinthal (1990, 131) makes it clear that, at least implicitly, they rely on concepts from that theory:

> Two related ideas are implicit in the notion that the ability to assimilate informa-
> tion is a function of the richness of the pre-existing knowledge structure: learning
> is cumulative, and learning performance is greatest when the object of learning is
> related to what is already known. As a result, learning is more difficult in novel
> domains, and, more generally, an individual's expertise – what he or she knows
> well – will change only incrementally. The above discussion also suggests that
> diversity of knowledge plays an important role. In a setting in which there is un-
> certainty about the knowledge domains from which potentially useful information
> may emerge, a diverse background provides a more robust basis for learning be-
> cause it increases the prospect that incoming information will relate to what is al-
> ready known. In addition to strengthening assimilative powers, knowledge *diversity*
> also facilitates the innovative process by enabling the individual to make *novel*
> *associatons* and *linkages* [emphasis added].

On the other hand, diversity may restrict the level of prior knowledge and
thus diminish 'assimilative powers' with regard to complex new information.
There is no doubt a trade-off between diversity and level of knowledge of an
individual; if diversity goes beyond a certain point, the level of knowledge per
field of knowledge will diminish.[5]

From an organizational point of view specialization has always been the
answer to problems of insufficient knowledge levels; by having specialists in
various domains (diversity) one can obtain a high level of knowledge on each
domain. There is, however, a limit to this solution too. The more the people
have a similar background, similar experience, education, knowledge at the
workfloor, the better they are able to communicate, but the less they can learn
from each other. By contrast, the more the knowledge of people differs, the
more they can potentially learn from each other, but the more difficult communication becomes. This trade-off between diversity and similarity (or commonality) is also recognized by Nyström (1979).

Note that we examined the effects of diversity at the industry level in
Chapter 5. In contrast to the treatment above the source of diversity was the
entire firm. The more firms that are present in the industry, the higher the
diversity of outlooks and opinions on technological opportunities. At this level

similarity can play a role too. In our discussions on technological cooperation we will focus on small steel firms in the US (the so-called 'minimills') that undertake quite similar incremental innovation projects. They communicate extensively (and apparently without many difficulties) about the results of these projects with each other.

In their discussion of the absorptive capacity of the organization Cohen and Levinthal (1990) state that this capacity depends on the absorptive capacity of its individual members. At the same time they stress that 'A firm's absorptive capacity is not . . . simply the sum of the absorptive capacities of its employees . . .' The way the firm organizes the scanning of the environment for external information, and organizes the selection, evaluation and use of the external information, is important in assessing its absorptive capacity. An important aspect are the communication structures of the firm with the environment and between the various subunits. Cohen and Levinthal (1990, 132) argue that the way these structures are organized depends on the expertise of the people in the firm and the nature of the external information. Communication structures can be more centralized or decentralized and dependent on people with more or less diverse backgrounds. In addition there can be less structured patterns of communication. With regard to centrality, they say (ibid., 132):

> The interface function may be diffused across individuals or be quite centralized. When the expertise of most individuals within the organization differs considerably from that of external actors who can provide useful information, some members of the group are likely to assume relatively centralized 'gatekeeping' or 'boundary-spanning' roles . . . For technical information that is difficult for internal staff to assimilate, a gatekeeper both monitors the environment and translates the technical information into a form understandable to the research group. In contrast, if external information is closely related to ongoing activity, then external information is readily assimilated and gatekeepers or boundary-spanners are not so necessary for translating information. Even in this setting, however, gatekeepers may emerge to the extent that such role specialization relieves others from having to monitor the environment.

Relating this to the earlier discussed photolithographic alignment equipment industry, one can imagine that for the core technology and each of the closely related component technologies a gatekeeper exists.

As said before, the scanning of the environment by the gatekeeper is one thing, sharing the knowledge with the research people is another thing. There must be enough 'background knowledge' in the firm in order to communicate with 'specialized' gatekeepers. This brings Cohen and Levinthal (1990, 134) to a similar sort of tradeoff as Nyström made:

The observation that the ideal knowledge structure for an organizational subunit should reflect only partially overlapping knowledge complemented by nonoverlapping diverse knowledge suggests an organizational trade-off between diversity and commonality of knowledge across individuals.

Partially overlapping knowledge structures are especially important for communication between subunits of the firm. With respect to innovation it is well known that the interfaces between R&D, production and marketing are very important for the ultimate use of new information and thus for effective technological learning. These interfaces are constituted by effective communication structures, direct personal contacts between functions, overlapping problem-solving cycles, and so on.

To sum up, the abilities of the gatekeeper determine the quality of information selection and translation from the environment, but sharing this information with the rest of the organization requires a certain overlap in the knowledge of the gatekeeper and the others. This brings up a specific point. When technological change in the environment is rapid and its direction uncertain, it may be risky to rely on strongly centralized information structures, for a gatekeeper may not know how and where in the firm a new piece of information is best applied. 'Under such circumstances, it is best for the organization to expose a fairly broad range of receptors to the environment' (ibid.,

Figure 7.2 Two stylized information structures of the firm

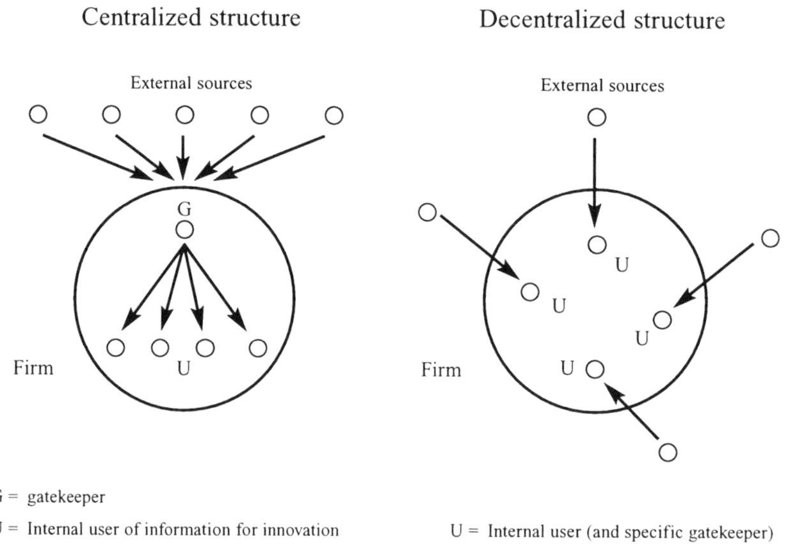

G = gatekeeper
U = Internal user of information for innovation   U = Internal user (and specific gatekeeper)

132). Figure 7.2 respresents a centralized and a decentralized information structure of the firm.

This discussion of absorptive capacity has made clear that the traditional notion of R&D expenditure does not completely reflect 'technological learning'. R&D expenditure to create new products and processes is one thing, proper communication structures of the R&D department(s) with the environment and with other subunits is another thing. At the same time, however, effective and efficient communication structures are guideposts to the firm. Through its absorptive capacity a firm may learn which kinds of technological development in the environment are important for the firm and subsequent R&D expenditure may follow. The absorptive capacity – as a reflection of the 'quality' of the R&D unit and the internal communication structures inside the firm – thus influences the external information perceived and acquired, as Figure 7.3 reflects.

Traditionally the economics of innovation recognizes technological opportunities, demand conditions, and appropriability conditions as factors determining R&D expenditure. Cohen and Levinthal (ibid., 140) make clear that absorptive capacity may be a kind of intermediary variable. The R&D intensity reflects the learning incentives of the firm, which are mediated by its absorptive capacity. Figure 7.4 represents the basic model of how absorptive capacity affects the determination of R&D expenditure.

Figure 7.3 Absorptive capacity and external information sources

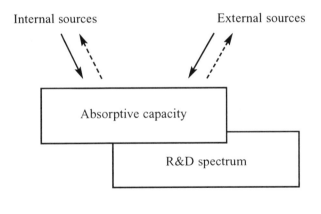

Internal sources   External sources

Absorptive capacity

R&D spectrum

⟶ Utilizating of sources (information flow)

---▶ Identification / perception of sources

The more there is to learn (technological opportunities), the more the firm will try to learn through R&D investment, but this depends on the firm's beliefs and expectations about technological opportunities (determined by its absorptive capacity). Matters are complicated because technological opportunities not only indicate the 'quantity' of learning, but also the complexity of learning. The nature of the technological opportunities determine the complexity of learning, and the more complex learning, the more R&D has to be accumulated to establish an absorptive capacity with which the learning 'needs' can be appreciated.[6]

Appropriability conditions (the existence of spillovers; see below) depend on the effects of spillovers on the firm's profits, represented by 'competitor interdependence' (which relates to the nature of demand). According to the model, the effect of appropriability conditions on R&D spending is also mediated by the absorptive capacity. This could be interpreted as the firm's ability to select innovation projects which are difficult for rivals to learn about.[7]

In general, the R&D spending in the past influences the absorptive capacity of the firm which, in its turn, determines the incentive to learn, given appropriability and technological opportunities. Normally, the perceived needs

Figure 7.4 Model of absorptive capacity and R&D incentives

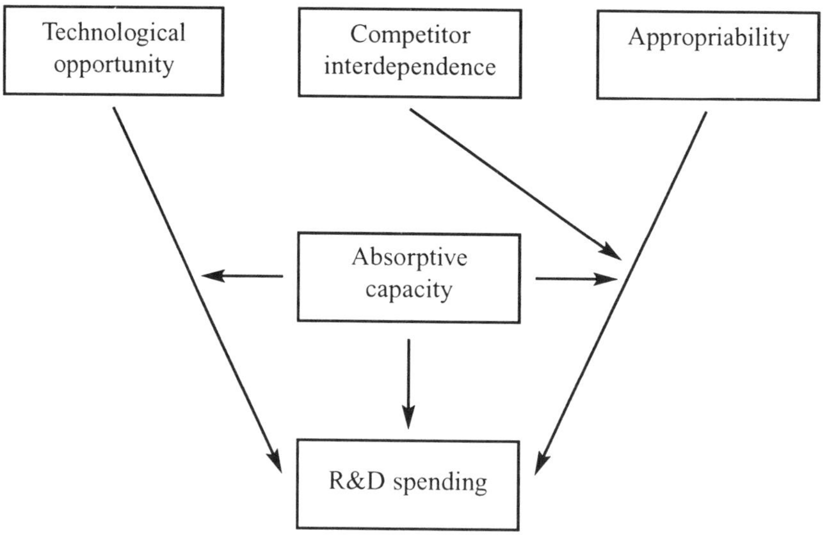

Source: Cohen and Levinthal (1990, 140)

to learn causes additional R&D investment. In the extreme case when external knowledge can be assimilated (turned into innovations) completely without any new expertise, additional R&D investments are not required.[8]

Spillovers

When we described the problem of this chapter it was said that knowledge is both input to and output from the innovation process. A related point is that R&D for a specific innovation project (equipment, but especially 'human capital') should not be seen in isolation. A currently new project P_{it} of firm i is probably related to a previous project ($P_{i(t-1)}$), to another current project P'_{it} or to a project of another firm, now or from the past (P_{jt} or $P_{j(t-1)}$). Innovation is concerned with accumulation of knowledge, with years of R&D expenditure, and with the build-up of specialized knowledge which cannot be acquired quickly by hiring people from outside. The accumulation is firm-specific and information from inside and outside sources is painstakingly added to the 'stock of knowledge' (if at all useful). When an innovation project from firm i profits somehow from the knowledge build up in another innovation project from firm j, we speak of knowledge spillovers. Orginally the whole issue of spillovers was related to the appropriability problem; if part of the, or the entire, knowledge from a specific innovation project of a firm spills over to one or more other firms, the first firm runs the risk of not being able to receive all or a large part of the potential profits from its newly created knowledge. But there are, of course, two sides of the coin. Our firm perhaps could not have made the innovation without the knowledge spilled over from other firms' innovation projects.

If we limit the analysis to spillovers between firms in the same market, then two possibilities are distinguished: substitution and complementarity. The substitution effect of technological spillovers occurs when results of R&D spilled over from firm j result in a reduction of the R&D expenditure of firm i. Firm j's R&D is substituted for one's own R&D. One speaks of complementarity when the spillover from firm j leads to an increase of the R&D from firm i (see, for example, Bernstein, 1989; Bernstein and Nadiri (1988); and Levin and Reiss, 1988).

Learning from other firms

In Chapter 3 transactions cost economics has been discussed as a possible economic theoretical framework to explain the organization of R&D, that is, innovation activity. In principle various parts of innovation activities of one or more projects could be undertaken by different firms. In the logic of transaction cost economics this would imply that one firm is the main contractor and that it sources out some parts of the innovative activities. A well known example is the sourcing out by pharmaceutical firms of the 'clinical' part of R&D.[9]

More generally it can be stated that a division of labor in R&D, resulting in a number of firms carrying out different parts of a product development project, implies that one firm learns from the other.

In a regime of cumulative technology an innovating firm learns from its own previously completed projects. One could add existing customers and suppliers to this picture. As was the case with intra-organizational learning (Henderson and Clark, 1990), the learning of a manufacturer and its most advanced suppliers and customers possibly evolves along a given set of communication channels and problem-solving procedures. In contrast, radical innovations in a regime of discrete technology usually require breaking out of 'settled structures'. As we suggested before, many radical innovations come from firms or institutes outside an industry. Furthermore, more distant information sources, such as foreign research institutes, are used by radical innovators, compared to incremental innovators.

4   CONCLUSIONS

In this chapter the creation of innovations is seen as a learning process. What is learned is the knowledge incorporated in new products, new processes, and results from technology projects. The learning concept must, however, include 'learning how to innovate'. Learning effective procedures to create innovations and to exploit them economically are as important as technological problem solving and creativity.

Technological problem solving and creativity were already discussed in Chapter 4. The concept of creativity of Usher emphasized the 'learning what' and the individual. The creative individual operates in an institutional context. This chapter made clear that individuals can learn from inside as well as outside sources. Learning from inside sources is related to communication patterns, that follow partly from the organizational structure of the firm. We have already discussed the impact of communication channels and problem solving procedures on the firm's ability to learn new things with the discussion of the concept of architectural innovations. It has become clear that such procedures and channels (learning how) in fact strongly influence what is learned by individuals and the organization as a whole.

The concept of absorptive capacity made clear that 'learning what' at the level of individuals is dependent on what has already been learned. In relation to 'learning how' the concept can be interpreted as learning by the organization as a whole through the proper conditions offered to learning individuals. An important aspect is the (informal) communication patterns that enhance mutual learning by diverse people. There is a limit to learning from each other on the grounds of diversity. A certain degree of commonality between individuals is needed to be able to understand one another.

NOTES

1. For the time being we speak of 'knowledge', but it will become clear that in order to innovate new information is transferred to various people involved in the innovation process. These people transform (learning) and transfer this information in many consequetive 'stages' (see the concept of Usher in Chapter 3).
2. One is reminded of the data concerning the development costs of nylon, where it became apparent that investment in new production facilities accounted for the majority of costs to bring it on the market. Design activities may be a source of confusion here. In the management of innovation literature 'design' is often used for describing the 'prototype', which is withou any doubt a part of the R&D activities. In the study of Napolitano 'design' may be exclusively used to indicate the esthetic design of the product, which is sometimes undertaken by a separate subconpe', which is withou any doubt a part of the R&D activities. In the study of Napolitano 'design' may be exclusively used to indicate the esthetic design of the product, which is sometimes undertaken by a separate subcontractor.
3. The score for this factor is on average 4.2 (from a maximum of 6), while the second most important source is 'design', with a score of 2.4).
4. If one looks at the definition of product innovation used in the study, it seems as if innovations that are created by a supplier can not be excluded completely: '...an innovative product is a product which could allow a firm to enter a new market, or a product substantially different, from a technological stand-point, from any other product previously manufactured by that firm' (Napolitano, 1991, 172).
5. Related to this is the question of the optimal number of innovation projects a person is simultaneously involved. According to Wheelwright and Clarke (1992) the effectiveness of engineers and researchers (and thus the effective use of their problem-solving ability) diminishes somewhat after two to three projects. The main reason in this case is the time needed for managing and coordinating tasks for the different projects. Purely from a cognitve point of view, the optimal number of projects may be a bit higher.
6. In my opinion, this implies that R&D spending also determines absorptive capacity.
7. Cohen and Levinthal (1990) themselves do not give an explanation of the mediating role of absorptive capacity in the effect of appropriability on R&D.
8. In the case of the Italian manufacturing industry we saw that many firms innovate, while they carry out no R&D. To the extent that this is not caused by statistical problems in measuring all innovative activity by using R&D expenditure, these firms can be said to have an absorptive capacity that has been established without any R&D spending.
9. The clinical part of R&D in the pharmaceutical industry concerns testing the new drug on animals (stage 1) and on people (stages 2 and 3). This testing of people is usually done in (large) hospitals.

REFERENCES

Anderson, J.R., R. Farrell and R. Sauers (1984), Learning to program LISP, *Cognitive Science* 8, 87–129

Bernstein, J.I. (1989), The structure of Canadian inter-industry R&D spillovers and the rate of return to R&D, *The Journal of Industrial Economics* 37, 3, 315–328

Bernstein, J.I. and M.I. Nadiri (1988), Interindustry R&D spillovers, rates of return, and production in high-tech industries, *AEA Papers and Proceedings* 88, 2, 429–34

Brouwer, E. and A. Kleinknecht (1994), Innovatie in de Nederlandse industrie en dienstverlening (1992), Report for the Ministry of Economic Affairs, The Hague

Chesnais, F. (1996) Technological agreements, networks and selected issues in economic theory, in R. Coombs et al., *Technological Collaboration*, 18–33

Cohen, W.M. and Levinthal, D.A. (1989), Innovation and learning: the two faces of R&D, *Economic Journal* 99, 569–96

Cohen, W.M. and D.A. Levinthal (1990), Absorptive capacity: a new perspective on learning and innovation, *Administrative Science Quarterly* 35, 128–52

Coombs, R., A. Richards, P.P. Saviotti and V. Walsh (eds) (1996), *Technological Collaboration: The Dynamics of Cooperation in Industrial Innovation*, Cheltenham, Edward Elgar

Docter, J. and C. Stokman (1987), Innovation in manufacturing medium-size and small enterprise: knowledge breeds prospects, in R. Rothwell and J. Bessant (eds), *Innovation: Adaptation and Growth*, Amsterdam, Elsevier, 213–26

Dodgson, M. (1993), Organizational learning: a review of some literatures, *Organization Studies* 14, 3, 375–94

Dodgson, M. (1996), Learning, trust, and inter-firm technological linkages: some theoretical associations, in R. Coombs et al., Technological Collaboration, Cheltenham, Edward Elgar, 54–75

Gibbons, M. and R. Johnstone (1974), The roles of science in technological innovation, *Research Policy* 3, 220–42

Henderson, R. and K. Clarke (1990), Architectural innovation: the reconfiguration of existing product technologies and the failure of established firms, *Administrative Science Quarterly*, 35, 9–30

Jewkes, J., D. Sawer and R. Stillerman (1969), *The Sources of Invention*, London, Macmillan

Kogut, B. and U. Zander (1992), Knowledge of the firm, combinative capabilities, and the replication of technology, *Organization Science* 3, 383–97

Levin, R.C. and P.C. Reiss (1988), Cost-reducing and demand-creating R&D with spillovers, *Rand Journal of Economics* 19, 4, 538–56

Monteverde, K. and David J. Teece (1982), Supplier switching costs and vertical integration in the automobile industry, *Bell Journal of Economics* 12, , 206–13

Napolitano, G. (1991), Industrial research and sources of innovation: a cross-industry analysis of Italian manufacturing firms, *Research Policy* 20, 171–78

Nelson, R.R. and S.D. Winter (1982), *An Evolutionary Theory of Economic Change*, New York, Belknap Press

Nyström, H. (1979), *Creativity and Innovation*, New York and London, Wiley

Pavitt, K. (1987), The objectives of technology policy, *Science and Public Policy* 14, 4, 182–8

Senker, J. and W. Faulkner (1996), Networks, tacit knowledge and innovation, in R. Coombs et al., *Technological Collaboration*, 76–97

Williamson, O.E. (1985), *The Economic Institutions of Capitalism*, New York, The Free Press

Wheelwright, S.C. and K.B. Clark (1992), *Revolutionizing Product Development; Quantum Leaps in Speed, Efficiency, and Quality*, New York, Free Press

8. Private patterns of technological cooperation

1 INTRODUCTION

In the first part of the book several problems have been discussed with which innovators are confronted. Some of these problems can, at least under certain circumstances, be solved by cooperation with other private firms. Technological cooperation between private firms is the topic of this and the following chapters. It not only relates to innovation problems mentioned in the earlier chapters, such as the financing or the speed of R&D, but also to the learning issue in the previous chapter. The related question in the latter case is 'how can two or more firms that form a technological cooperation learn from each other?'. In the former case it is about the possibility of reducing the costs or increasing the speed of innovation through cooperation.

The topic is divided into two parts. In the empirical part of this chapter a general description is given of the various forms of cooperation in different industries. It is about general patterns of cooperation. The theoretical part addresses the questions of motives for cooperation and the choice of specific forms of cooperation. In the next two chapters we go into more detail by describing specific cases of cooperation. Chapter 9 is concerned with cooperation between manufacturers and suppliers. Chapter 10 focuses on cooperation between firms in the same industry.

The general pattern of cooperation in this chapter is explained by focusing on economic and other theories that identify the conditions in the environment under which cooperation is expected to give better results than no cooperation. Such theories usually do not explain some of the organizational designs encountered in practice (and described in the next chapters), nor the selection of a specific partner. Theories related to these aspects are discussed in the theoretical section of Chapter 9.

The problem

Economics in general is not interested in explaining why one firm performs better than another. It wants to know what kinds of general patterns of behavior and performance can be perceived within a group of firms. The 'dominant group' in theoretical approaches to economics is the market. A market is economically defined in terms of a high degree of substitution between rival products, that results in an (over)emphasis on competition between firms in a particular market. Cooperation traditionally had a negative connotation: 'collusion' between firms in a market is a purposeful attempt to increase profits at the cost of the consumers.

Empirically cooperation is not a new phenomenon. Cooperatives have existed for many decades, especially in agriculture. Usually farmers do not perceive colleages in their region as competitors; they all produce for anonymous customers. In the manufacturing industry many employer associations aim at defending the common interests of the firms in a particular industry, for instance through lobbying in government circles. They also provide their members with technical and commercial information about new equipment and the like. Cooperation between firms on issues in which they formerly competed seems to be rather new, or at least a topic not studied systematically by economists. Apart from well-documented cases of cartels and other forms of collusion, mainly anecdotal evidence exists that cooperation between firms in an industry occasionally occurred in the period 1850–1970. So when technological cooperation, such as developing a new product together, occurred more frequently, it immediately became a puzzle for economic theory. Theories on technological cooperation probably are still underdeveloped.

In Chapters 3 to 5 we learned about factors inhibiting the firm's innovation process. Theoretically these factors can be summarized under the headings of technological uncertainty and market uncertainty. Both types of uncertainty are assumed to increase in most markets today. Technological uncertainty increases because the speed of innovation increases, while the complexity and the number of technologies needed for innovation increases too. Market uncertainty increases with the internationalization of markets. In regional, 'protected' markets the number of competitors is normally less, they are better known, and the structure and nature of demand is more transparant (local customers). Under pressure of an increasing competition one may assume that firms try to reduce uncertainty.

In Chapter 6 'public' initiatives to reduces market and technological uncertainty for firms have been discussed. Here the focus is on 'private' initiatives. A way for a private firm to reduce technological uncertainty is by cooperating with other firms. Chapter 7 showed that a firm can learn from outside sources of information. One main motive for technological cooperation between private firms is to learn from the other firm(s). A basic logic behind this idea of cooperation is that 'two know more than one' or can do more (in terms of expenditure on R&D). Market uncertainty for a firm follows from the imitation danger of competitors or from the danger that one or more of them innovate earlier. Cooperation may also reduce market uncertainty as the number of competitors 'which may surprise you' will be reduced. The main problem is that no matter how promising some forms of technological cooperation may seem, they have a cost: every form of cooperation reduces the profits from innovation compared to the situation that a single firm successfully completes an innovation project and exploits its commercial potential completely. In addition a firm cooperating with others may see the (agreed upon) sharing of innovation profits as problematic. In short, technological cooperation may reduce market or technological uncertainty, but potential monopoly profits are reduced too. In situations where innovations are created that each firm separately would not have been able to create, an unequal spread of costs and benefits may undermine the cooperation. In the empirical part of the chapter we look at the kinds of cooperation in practice. In the theoretical part we look at how different forms of cooperation are possibly used for different purposes. More generally we examine the motives for cooperation and we relate specific motives to specific organizational forms.

2 THE EMPIRICAL SIDE

We define technological cooperation of a firm as a formal or informal agreement with one or more other firms, in which the firm receives technological information that can be used to enlarge its knowledge base, develop new technologies, and develop or improve new products and processes. In return the firm offers the other firm(s) something, in many cases also some technological information. This definition includes agreements in which one firm buys a 'piece' of technological knowledge from another firm, that is clearly defined and bounded ex ante. An example is the purchase of technological knowledge through a license contract. In such a purchase of technology the technological knowledge is already fully developed, it is merely passed on to other hands. We will not deal with this transfer of technology that has already been used by others. Instead the emphasis is on the development of new technology and on the transfer of technological information that cannot be readily used. In the latter case the information is used as an input to the innovation process.

The definition given above can be seen as 'strategically underdeveloped' and one-sided. The emphasis lies on the cognitive aspect of cooperation, the transfer of technological knowledge to the firm under consideration. A more strategic definition of technological cooperation is as follows. A strategic technological alliance is a form of cooperation in which two or more firms set common goals with respect to technology and/or product development and act accordingly.

We argued before that technological cooperation is not a new phenomenon. During the interwar period cooperation between large internationally operating firms was well established. Cross-licensing of patents was quite common and also an extensive sharing of scientific and engineering information took place between the large firms (Cantwell and Barrera, 1995).

According to several authors technological cooperation has been increasing considerably in the last two decades. Many of these empirical studies on technological cooperation are based on a limited set of firms. Brouwer and Kleinknecht (1994) state that their study is the first to encompass the whole manufacturing sector and the service sectors in a country, in this case the Netherlands. Table 8.1 gives an illustration of the importance of technological cooperation in terms of the percentage of Dutch firms who responded in the related questionnaire that they had some form of technological cooperation in 1992.

Obviously, one could question whether the data on cooperation in the Netherlands are representative for the situation in most industrialized countries. One feature of the Dutch economy is a lack of large firms. Except for the presence of a number of large multinationals, relatively few large firms exist. This cannot be extracted from Table 8.1, where all firms bigger than 500 employees are classified in one category. Figure 8.1 provides some data on

Table 8.1 The most important types of partner in technological cooperation in the Dutch economy in 1992, according to size class of firms

Size class	Some	Customers	Suppliers	Competitors	Universities	n
Manufacturing						
10–19	28	5	11	0	18	26
20–49	36	18	13	4	7	91
50–99	40	21	14	5	14	229
100–199	50	22	23	4	17	195
200–499	59	30	39	8	22	164
>500	82	48	47	22	43	102
Total	**43**	**21**	**19**	**5**	**15**	**807**
Services						
10–19	41	6	11	7	13	19
20–49	55	23	9	19	22	34
50–99	46	22	22	9	15	55
100–199	63	24	15	13	20	37
200–499	50	28	26	13	22	88
>500	64	29	43	15	26	91
Total	**50**	**19**	**16**	**12**	**18**	**324**
Overall	**47**	**20**	**17**	**9**	**17**	**1.131**

Source: Derived from Brouwer and Kleinknecht (1994, 7)

Figure 8.1 Technological cooperation in the 25 largest firms in several European countries

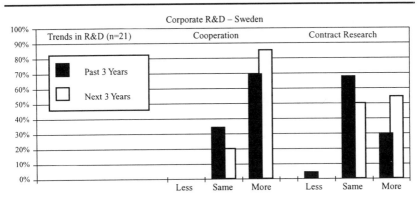

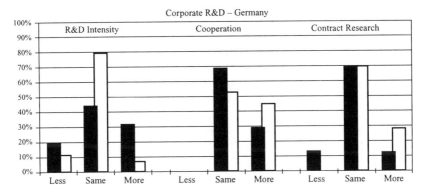

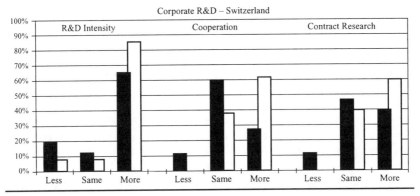

Source: Various consultancy reports on technological cooperation written for the Ministry of Economic Affairs in the Netherlands

Figure 8.2 Forms of cooperation according to direction, intensity and autonomy

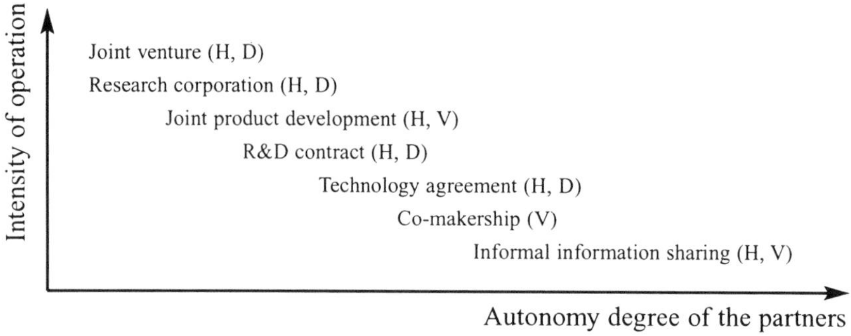

Intensity of operation ↑

Joint venture (H, D)

Research corporation (H, D)

Joint product development (H, V)

R&D contract (H, D)

Technology agreement (H, D)

Co-makership (V)

Informal information sharing (H, V)

→ Autonomy degree of the partners

H= Horizontal, V= Vertical, D= Diagonal

technological cooperation of the largest firms in several European countries. Although this figure does not show whether the 'largest' firms cooperate more than 'larger' firms (>500), it does distinguish cooperation from contract research.

The purchase of new technology for a specifically defined use, that was included in our definition of cooperation at the beginning of this chapter, must be excluded from 'contract research' as reflected in Figure 8.1. Both types of cooperation seem to grow in importance, with contract research possibly slightly less than technological cooperation as such.

Hagedoorn and Schakenraad (1990a) give an account of a much larger set of technological cooperations in all industrialized countries, but confined to a limited number of industries. Table 8.2 illustrates the pervasiveness of technological cooperation in the information technology industry, which encompasses a large number of 'hightech' markets. These IT markets are characterized by strong international competition, but the worldwide 'network' of technological cooperation suggests that some mixture of competition and cooperation exists. To understand better this co-existence of competition and cooperation, one needs to know what kinds of cooperation are involved in the IT industry and the other industries included in the database from which the study draws.

Forms of technological cooperation can be classified according to 'direction' and 'intensity'. The direction can be vertical, horizontal, and diagonal (or 'other'). Vertical cooperation is between a firm and its suppliers ('upstream') or its customers, including distributors ('downstream'). In the case of technological cooperation downstream is usually confined to firms which purchase goods or services from other firms. Intensity relates to the intensity of cooperation, which is usually defined in terms of the degree of autonomy of the

Table 8.2 Forms of technological cooperation

	Total		Bio-technology		Information technology	
	Number	**%**	**Number**	**%**	**Number**	**%**
Joint ventures	790	26	219	22	571	28
Exchange agr.	1272	42	365	37	907	44
X-licensing	53	2	10	1	43	2
Second-sourcing	120	4	1	0	119	6
Minority stakes	601	20	234	24	367	18
R&D contracts	185	6	151	16	34	2
Total	**3021**	**100**	**980**	**100**	**2041**	**100**

Source: Hagedoorn and Schakenraad (1990a, 178)

cooperating partners. Figure 8.2 shows a classification of cooperation according to direction and intensity.

Especially with regard to technological cooperation this classification according to intensity has its limits. A loss of autonomy of cooperating firms does not necessarily imply that these firms cooperate intensively in terms of technological information exchange and vice versa. A loose agreement on technology sharing by a number of firms may involve an intense exchange of technological information, depending on the attitude of the people involved. The loss of autonomy in technological cooperation has to do with the extent of irreversibility of the investments made by the partners. Normally large investments by the partners would imply the intention to exchange relatively large amounts of information (if that is the main motive for the cooperation), but this need not necessarily be so.

The earlier mentioned study of Brouwer and Kleinknecht (1994) contains cooperation by firms according to 'main types of partner', for example, customers, suppliers, competitors, but does not reveal the specific kinds of agreement. In the manufacturing industry 43 percent of the firms with R&D or innovative activities say that they undertake technological cooperation, 21 percent have cooperation(s) with customers, 19 percent with suppliers, 5 percent with competitors, and 15 percent with universities. Technological cooperation in Dutch service sectors shows a similar pattern.[1]

Hagedoorn and Schakenraad (1990a) do classify the technological cooperations in biotechnology and information technology according to forms. Table 8.2 gives an overview.

In the table 'joint ventures' includes 'research corporations'. In both cases the cooperating partners start a separate common firm. Joint ventures are normally dealing with R&D, production, and marketing, while research corporations are more exclusively focused on specific research programmes. In both forms of cooperation profits and losses are shared according to equity investment. Technology 'exchange agreements' cover a number of agreements which regulate technology sharing and/or transfer between two or more companies, such as:

• technology sharing agreements;
• joint-development agreements;
• joint research pacts establishing joint research projects with shared resources. (ibid., 179–80).

The number of joint ventures as well as exchange agreements have been growing in recent years, while the other forms of agreement from Table 8.2 are not only reported less frequently, but also remained more stable during the years 1970–88 (ibid., 179). The database discussed here is about technological cooperations announced publicly. It is possible that some forms of cooperation are less publicly known than others. Cross-licensing is a special form of licensing where partners exchange information covered by patents. Minority stakes are usually taken by a large firm in one or more smaller firms which operate in technological fields outside its own core technology. In this way the firm has at least some access to the research results of these adjacent fields. Research contracts involve specific parts of the R&D of a firm contracted out to another firm. They take two forms. One is that a large firm uses a small 'research firm' (especially in biotechnology) for carrying out a specific project. The other is that a large firm contracts out to another large firm. In this case usually both firms offer each other a contract. Second sourcing is a type of technological cooperation where one party allows another, usually smaller, firm to undertake part of the production, by offering the specifications of product and process; this is a form of vertical cooperation dicussed in the next chapter. All the other forms discussed here are most of the time involving firms in the same industry or in unrelated industries ('diagonal'). According to Hagedoorn and Schakenraad (ibid.), research contracts, cross-licensing, and second sourcing occur more frequently than the figures suggest. For research contracts and cross-licensing firms do not have the habit announcing them publicly, secound sourcing has become so common that firms do not mention them frequently.[2]

Most empirical studies on technological cooperation among large groups of firms do not make a distinction between horizontal, vertical and diagonal cooperation. Today horizontal and vertical cooperation are both quite common, diagonal forms of cooperation are relatively rare. In Table 8.2 most of the cooperations mentioned are between firms in the same industry. The figures on cooperation of small and medium-sized firms will probably include a large number of cooperations with customers. In Chapter 9 we discuss the importance of vertical cooperation.

Wolek (1979) examined differences in the percentage of the total amount of R&D performed by cooperation between industries. Industries with a low R&D intensity in general have a higher portion of cooperations. Examples are the 'gas and electric' industry and the 'printing' industry. According to Mowery (1989) more recent trends in technological cooperation show that international alliances occur in mature industries with relatively low levels of R&D (steel, automobiles) and in 'young' industries (biotechnology) and industries with a high R&D intensity (commercial aircraft, robotics). In the MERIT database on technological cooperation 42 percent of all alliances are in information technology and a large majority are in this field, biotechnology and new materials (Hagedoorn and Schakenraad, 1990b).

If one takes (international) alliances and other forms of technological cooperation, including vertical relationships with customers and suppliers, no significant differences between industries occur, again with the exception of public utilities. This picture came out of a survey among the member countries of the EU (Kleinknecht, 1996).

3 THE THEORETICAL SIDE

There are very few economic theories which can give an explanation, even partially, for the strongly rising number and types of technological cooperation in the last two decades. The phenomenon of technological cooperation as such is not a main topic of economic theory. Some new theoretical developments occur around the issue, while a few existing theories include cooperation in their general aim of explaining economic organization. Today transaction cost economics is perhaps the most prominent of the latter class of theories. We discuss the explanation this theory provides for technological cooperation extensively in the next chapter, because in our opinion it is better suited to explain vertical cooperation than horizontal cooperation. In contrast, some other theoretical explanations are better suited for horizontal cooperation, so we discuss them in Chapter 10. Here we limit the discussion to the main features of transaction cost theory and the development of new theoretical elements specifically addressing the technological cooperation issue. We also include uncertainty reduction. In addition we analyze the motives for cooperation, which come largely from management studies. They provide us with some empirical backing for (economic) approaches to cooperation.

Separate attention is paid to an explanation why sometimes a more intense form and sometimes a looser form of cooperation is chosen.

Uncertainty reduction

For some firms the perceived technological uncertainty or market uncertainty is greater than for other firms. The technological uncertainty may differ due to a difference in experience with similar innovation projects in the past or to a difference in availability of technical expertise in certain areas. Market uncertainty may differ, because of a difference in perceived market power or in control over specific complementary assets. If a potential partner is perceived as having some assets which may reduce the firm's technological and/or market uncertainty of innovation, it may be decided to enter a form of cooperation. This suggests that there are technological motives and motives related to appropriability. And indeed, in the discussion of motives it will turn out that such a distinction is made.

Teece (1986), as discussed in Chapter 4, suggests that cooperation weakens the innovating firm's appropriability situation, unless activities are contracted out which are quite standard and thus not specific to the innovation. It must be stressed here that uncertainty relates to the innovation project which has yet to be undertaken. When a firm perceives a high market or technological uncertainty, cooperation may improve the probability of commercial success. In markets with rapid technological change, market and technological uncertainty are on average higher than in markets with a slower technological change. Teece (1992) argues that technological cooperation in markets with rapid technological change and increased competition can be explained because of the increased 'dispersion of technological competences'. Such forms of cooperation are mostly horizontal or diagonal, so this explanation is further discussed in Chapter 10.

Transaction cost economics

Hagedoorn (1993, 371) gives an appropriate description of the lack of theoretical explanations for technological cooperation and the particular use of transaction cost economics (TCE) in explaining it:

> So far a few theoretical contributions have been made that have put this subject on the agenda. In particular transaction cost economics inspired contributions . . . have theorized interfirm partnering as an economic phenomenon in between market transactions and hierarchies. In general these transaction cost economics inspired contributions appear to concentrate on vertical, customer–supplier relations and discuss them in rather general theoretical terms or analyze concrete developments in case studies from which generalizations are difficult to make.

TCE is about the coordination of transactions regarding a particular product or service and the coordination of transactions across geographically dispersed production units or across units of a firm which produce different goods (multiproduct firm). Restricted to a specific good TCE is concerned with vertical cooperation. When the broader coordination problem is concerned (multiproduct, multiregion) both vertical and horizontal 'transactions' may be relevant. The transaction cost question in this latter case is, under which conditions is cooperation with another firm more efficient than full integration of a series of transactions, like in multidivisional firms, holdings, or multinational enterprises? This line of research will not be followed here.[3]

We give a brief general introduction to TCE below and apply this theory to the explanation of vertical cooperation in the next chapter.

TCE describes the conditions under which cooperation is a more efficient form of coordination than pure market contracts or pure hierarchical forms of coordination. At the background of the analysis are the generally applicable conditions of bounded rationality and opportunistic behaviour, that concern the coordination of all transactions. More specific to the choice of a particular form of coordination are the three characteristics of the transaction: the degree of specificity of the investments needed to undertake a transaction (asset specificity); the frequency with which the transactions occur per time period; and the uncertainty of the behaviour of potential trading partners (including internal units of the firm). TCE predicts that when asset specificity, frequency, and uncertainty have intermediate values, cooperation is the cost-efficient solution for coordination of the specific transaction. What it cannot explain is why at some time a joint venture is chosen and at other times another 'hybrid' form, such as a technology agreement. In addition it can be said that an efficiency explanation of technological cooperation is probably explaining only partly the rise of technological cooperations of the last 20 years.

New theoretical developments on technological cooperation

In the previous chapter it was argued that resource creation has been a neglected area in economic theory. Chesnay (1996) states that in networks of firms concerned with technology or specialized intermediary products resource creation is especially important. This can only be when each firm in the network is able to learn from the others. This requires two conditions: first the firm's capacity to learn; and second, communication channels and a common language with the partner firms. In Chapter 7 the firm's capacity to learn has been examined. With regard to the second condition the 'social tissue' of the network strongly influences the effects of the firm's attempts to learn from each other (Lundvall, 1992). In Chapter 11 we discuss networks further. In Chapter 10 we will show that technological specialization of firms in industries plays an important role in understanding horizontal cooperation.

Motives for cooperation

Especially in the management literature a vast amount of motives for technological cooperation are given. Hagedoorn and Schakenraad (1990a) give an overview. From this overview it can be derived that there are conditions under which technological cooperation occurs and there are specific motives. Conditions are, among other things, that partners must be of comparable technological sophistication, that the partnership is about applied research, and that the projects in the cooperation are not so large that when they fail they are a threat to the survival of each of the firms.[4]

Motives of small firms to cooperate may differ from those of large firms. Small firms usually lack the financial resources to keep up with 'state of the art' technological developments in more than one or a few fields. Technological cooperation for these firms is a must to keep a 'mimimum technological scope' to compete internationally. According to Hagedoorn and Schakenraad (ibid.,175) large firms prefer in-house development or the acquisition of firms when it regards their core technology, but:

> When the speed or uncertainty of particular fields of technology is extremely high and there is no time to develop know-how independently or when take-overs are too expensive, cooperative action is considered to be a realistic alternative. When co-operation is subsequently deemed necessary, it is frequently focused at:
>
> * the absorption of the partner's knowledge or technology;
> * the development of joint R&D and technological capabilities that are complementary to the companies' abilities;
> * expanding the companies' product range and future markets.

Hagedoorn (1993) covers an even larger set of studies dealing with motives for technological cooperation. On the basis of those studies, he makes a classification of motives into three main classes: (1) motives related to basic and applied research and technological development in general; (2) motives related to some concrete innovation processes; (3) motives related to market access and search for market opportunities. All the types of motives mentioned in the empirical studies are grouped according to these three main motives. Table 8.3 gives an overview.

At a more global level Table 8.3 shows the distinction between 'market motives' for technological cooperation and 'technological motives'. Especially the technological motives seem to relate strongly to the underlying forces of technological uncertainty, discussed previously. On the base of this list of

Table 8.3 An overview of motives for technological cooperation

1. Motives related to basic and applied research and some general characteristics of technological development:
   • Increased complexity and intersectoral nature of new technology, cross-fertilization of scientific disciplines and fields of technology, monitoring of evolution of technologies, technological synergies, access to scientific knowledge or to complementary technology;
   • Reduction, minimizing and sharing of uncertainty in R&D;
   • Reduction and sharing of costs of R&D;

2. Motives related to concrete innovation processes:
   • Capturing of partner's tacit knowledge of technology, technology transfer, technological leapfrogging;
   • Shortening of product life cycle, reducing the period between invention and market introduction;

3. Motives related to market access and search for opportunities:
   • Monitoring of environmental changes and opportunities
   • Internationalization, globalization and entry to foreign markets;
   • New products and markets, market entry, expansion of product range.

Source: Hagedoorn and Schakenraad (1990a, 178); the list of references to empirical studies for each 'type of motive' in the original table has been left out here.

motives from the literature, Hagedoorn (ibid.) lists around 4,200 technological cooperations from the MERIT data base on 10,000 cooperations according to industry and kind of motive. The results are presented in Table 8.4.

The technology/market (T/M) ratio is defined as the $\log^{10}$ of the ratio between a firm's total number of prevailing research inclined technological cooperations and its total number of predominantly market-related technological cooperations.[5] On the base of the results presented in Table 8.4 Hagedoorn (ibid., 32) concludes:

> it becomes obvious that, if one looks at the general outcomes, only three motives play a role of true significance. Technology complementarity, reduction of the innovation time-span, and market access and influencing the market structure are the most mentioned motives . . . Other motives appear to play only a very limited role.

From these figures it can further be interpreted that the two technological motives seems to play a more important role in hightech industries and fields than in mature ones. With the exception of telecommunication, computers,

Table 8.4 *Motives for strategic alliances, sectors and fields of technology*

	Number of alliances	Technol. Complement.	Basic R&D	High cost/ risks	Lack of financial resources	Reduction innovation time span	Monitoring technology market entry	Market access structure	Technol. market ratio
Biotechnology	847	35%	10%	1%	13%	31%	15%	13%	0.55
New materials technology	430	38%	11%	1%	3%	32%	16%	31%	0.15
Computers	198	28%	2%	1%	2%	22%	10%	51%	-0.18
Industrial automation	278	41%	4%	0%	3%	32%	7%	31%	0.18
Microelectronics	383	33%	5%	3%	3%	33%	6%	52%	-0.15
Software	344	38%	2%	1%	4%	36%	11%	24%	0.26
Telecommunications	366	28%	1%	11%	2%	28%	16%	35%	-0.04
Other I.T.	91	29%	2%	1%	0%	28%	24%	35%	-0.04
Automotive	205	27%	2%	4%	2%	22%	4%	52%	-0.25
Aviation/defense	228	34%	0%	36%	1%	26%	8%	13%	0.43
Chemicals	410	16%	1%	7%	1%	13%	8%	51%	-0.45
Consumer electronics	58	19%	0%	2%	0%	19%	9%	53%	-0.38
Food and beverage	42	17%	0%	1%	0%	10%	7%	43%	-0.35
Heavy electric/power	141	31%	4%	36%	1%	10%	11%	23%	0.17
Instruments/medical technology	95	35%	2%	0%	4%	40%	10%	28%	0.15
Other	76	9%	0%	35%	0%	6%	8%	23%	-0.41
Total	**4192**	**31%**	**5%**	**6%**	**4%**	**28%**	**11%**	**32%**	**0.06**

Source: Hagedoorn (1993, 27)

and microelectronics all hightech sectors indeed have a majority of their technological cooperations motivated by technological considerations, while in mature industries market motives prevail. The analysis is taken one step further by looking at the relationship between modes of cooperation and market respectively technology motive. It is concluded that joint ventures and other 'intense' modes of cooperation are used when a combination exists of underlying market and technological motives. Less intense modes of cooperation, like technology agreements, are more exclusively based on technological motives. The argument is that more intense modes require more (and more irreversible) investments, that are only undertaken when the cooperation has a number of commercial goals, covering both technology and market. Less intense modes have a one-dimensional goal, namely technological motives. Then communication occurs without considerable financial and organizational investments (that would have given both parties a perception of more control and perhaps more commitment).

Hagedoorn focuses on technological cooperation within sectors of the manufacturing industry, hence on horizontal cooperation. As said 'loose' forms of cooperation suffice when the parties focus on technology. The advantages can be divided into two classes: 'learning from each other' and 'cost reduction'. The last category includes reaping economies of scale in R&D. Economies of scale in R&D is not a well-documented topic. It is clear that in some industries innovation requires large-scale test facilities, and in process industries like chemistry, 'up-scaling' of the process to make the new chemical entity on a commercial scale can be very expensive. The larger the test facility that can be used by 'all' innovation projects of a firm, the lower will be the unit cost of the use of the facility by a project team. And the larger the full-scale process the more the innovation costs can be spread over a higher volume of output. Consequently if the test facility or scaling-up process can be enlarged by cooperation, economies of scale may follow. In many other industries it remains unclear to what extent single firms can gain unit cost advantages by undertaking the R&D jointly with one or more other firms. Moreover it can be argued that two firms can only profit from economies of scale when the joint research is really located at the same place. This would require investment in housing and equipment. It certainly means that profiting from economies of scale implies a level of investment of the parties that makes the 'loose' form of cooperation relatively intense. Another matter is a reduction of the total costs of R&D by cooperating with firms in similar technological areas. Logically the cost of R&D for each firm in a partnership decreases when the total cost of the joint project(s) is not, or not much, higher than when the firm would have undertaken the project(s) on its own. For many smaller firms horizontal cooperation is partially motivated by this possibility of cost reduction. In Chapter 10 the example of the MCC corporation is about cost reduction for large firms.

Market motives are about market penetration or the exercise of market power. In that case technological know-how is often exchanged for an access to or control of a market. Hagedoorn argues that cooperation based on market motives requires more intensive forms of cooperation than cooperation based on technological motives. Joint ventures often combine market motives and technological motives, because the range of activities includes product development as well as the production and marketing of the new product. As the nylon case at the beginning of the book made clear, production requires larger investments than R&D. Firms therefore seek organizational forms where they can, when needed, to control what goes on. The more intense forms of cooperation provide this control. Cooperations in which merely technological motives play a role can do with contractual modes where less control is possible.

In order to test these hypotheses the database on technological cooperation was split between strategic partnerships and non-strategic partnerships. The last part contains the majority of customer–supplier relationships and unilateral technology flows (for example technology licenses). With the exception of strategically oriented customer–supplier relationships the hypotheses were tested with data on horizontal (and diagonal) cooperations. Table 8.5 shows that a significant difference exists between the motives underlying complex modes and contractual modes of cooperation. Complex modes are indeed used more for a combination of technological and market motives than contractual arrangements.

In Chapter 10 two cases will be discussed. One research joint venture (the MCC Corporation) and the case of the minimills in the steel industry. The former can be seen as an intense form of cooperation, the latter as a loose form.

The analysis of organizational forms of cooperation related to market and technological motives does not deal with success or failure of technological cooperation. It can be assumed that when indeed a significant correlation

Table 8.5 T/M ratios of complex and contractual modes of strategic technology partnering

	Frequency of research-related motives (1)	Frequency of market-related motives (2)	1:2	T/ M ratio ($\log^{10}$)
Complex modes	601	863	0.696	-0.16
Contractual arrangements	1841	488	3.773	0.58

Source: Hagedoorn (1993, 36)

exists between organizational form of and motive for cooperation one form will not be consistently more successful than the other. But it may well be that the success rate of contractual modes of technological cooperation as well as joint-ventures and other forms of intense cooperation is determined by underlying organizational and managerial factors. Indications exist that cooperation between 'equal firms' is less stable than cooperation between 'unequal' partners. And when 'equals' cooperate the ones where a wide range of activities from product development towards marketing is undertaken, and where the joint venture has an autonomous management structure, are the most successful (Mowery, 1989, 27).

Cooperation in different industries

In the first part of the book Pavitt's typology of industries according to technological characteristics has been discussed. This typology is a good starting point to examine whether the need for, or possibilities of, technological cooperation differ per industry. In 'supplier-dominated' industries most innovations used by the firms are created by suppliers. An example is the textile industry where new machines are made by specialized machinery firms. in the last few decades the suppliers of information technology equipment and software have had a larger impact on the industry. On the basis of computer-aided design software some parts of the textile and clothing industries in high-income countries survived the major relocation of the industry towards low-wage countries. Although the emphasis is on innovative suppliers the pattern of production and use of innovations makes clear that technological cooperation between the firms in these industries and the dominant suppliers may exist. In the same way one could argue that the direction of cooperation in science-based industries is towards cooperation with scientific institutions. Pavitt also indicates a pattern of concentric technological diversification.

Von Hippel (1988) emphasizes the role of customers in the innovation processes in industries.[6] When customers in a single industry traditionally are the users of an industry's innovations it may occur that more and more customers are developing these machines or components/subsystems themselves or in cooperation with the suppliers. More generally some industries are characterized by a high percentage of innovations developed by the firms in the industry themselves. Examples are 'engineering plastics', 'tractor shovel-related products', and 'plastics additives' in which more than 90 percent of the innovations are realized by the manufacturers themselves. In other industries users (scientific instruments – 77 percent, semiconductors – 67 percent, poltrusion process – 90 percent) or suppliers (wire termination equipment – 56 percent) undertake the majority of innovations for use by those industries.

From the concepts of authors like Pavitt (1984) and von Hippel (1988) it can be derived that the direction of technological cooperation may differ per industry. Innovations are likely to be developed there where they give most profits. So when traditionally innovations in machines for the automobile industry were made by specialized machine suppliers to that industry, a change in market conditions could result in a situation where it becomes more profitable for the automobile manufacturers to make the new machines themselves. It may also occur that market conditions change in such a way that it becomes profitable for both parties to develop the machines jointly. Both explanations tend to put the locus of innovations for an industry to one side: the firms in the industry themselves, suppliers, customers, or the scientific community. Shifts from one to the other arise, but the rise of technological cooperation in most industries cannot be explained. This is not surprising when one notes that 'The structure of collaboration, the activities incorporated into collaborative ventures, and the amount and direction of technology transfer between' [the] firms all vary across industries, (Mowery, 1989, 19). Still a number of factors may have caused the relative advantage of cooperation above market transaction in technology (such as licenses) and corporate governance (cooperation between units within the firm). Relative to market transactions (intense) forms of technological cooperation allow for monitoring of the effectiveness of the intended transfer of knowledge or technology. They also make possible the pooling of noncodified, 'inseparable' firm-specific assets of various parties. This is all even easier within conglomerations, but cooperations provide a more rapid and less costly means to gain access to new technologies (Mowery, ibid., 23). Especially when technological change in the industry is rapid and firms are specializing more and more in subdomains of the whole set of technologies with which new products can be made (Teece, 1992). An increase in the cost of R&D and the shortening of product life cycles in many industries also make rapid market penetration of paramount importance (Mowery, ibid., 25). So access to new technology is the main factor in industries where new technologies occur and intraindustry specialization takes place. If in addition R&D costs rise strongly and product lifecycles become drastically shorter, market penetration motives become important too.

4 CONCLUSIONS

Although technological cooperation is not a completely new phenomenon it became clear from empirical studies that the number of technological cooperative agreements has been growing rapidly between 1970 and 1990. A more detailed analysis of the patterns showed that joint ventures and exchange (or technology) agreements are the most common forms of cooperation. Joint ventures are an example of an intense form of cooperation, technology agreements are much looser. Apart from intensity of cooperation a distinction has been made between horizontal, vertical, and diagonal forms. These will be studied further in the next chapters. This chapters revealed that more intense forms of cooperation combine technological and market motives, while less intense forms are limited to technological motives.

The extent of cooperation seems to differ between industries. On the one hand strongly regulated (public utilities) and some mature industries (steel, automobile) show a relatively large number of cooperations. On the other hand new industries and industries with rapid technological change are charaterized by extensive cooperation. A first theoretical explanation suggests increased competition, the shortening of product life cycles and the rising costs of R&D as important conditions for technological cooperation. All conditions are at work in industries with rapid technological change. In some mature industries strong competition and the related shortening of product life cycles form conditions for cooperation; market uncertainty is thus reduced. In new industries technological uncertainty is very high which suggests that a reduction of this uncertainty may be the main motive for cooperation.

NOTES

1. The data show quite clearly that the larger the firms are (up to the last size class of >500 employees') the higher the chance that they have one or more cooperative agreements with each of the types of partner.
2. In addition it can be added here that second sourcing may not be seen by firms as a form of technological cooperation. The next chapter will make clear that 'jobbers' are supply firms which merely produce on complete specification of the 'larger' firm, while 'main-suppliers' undertake their 'autonomous' product development in close contact with the large customer.
3. As far as I know, it has only been explained in TCE-terms why conglomerates and the like are more efficient than a series of market transactions. Whether cooperation may under certain conditions be more efficient than conglomerates has only been analyzed for single transactions.
4. Hagedoorn and Schakenraad refer to Haklisch (1986).
5. The ratio takes the value of zero, when both types of cooperation are equally frequent. A positive value means an inclination towards technological motives, a negative value towards market motives.
6. See table 2.5 in chapter 2 of von Hippel (1988).

REFERENCES

Brouwer, E. and A. Kleinknecht (1994), Innovation in the Dutch manafacturing and service sectors (in Dutch) (1992), Report for the Ministry of Economic Affairs, The Hague

Cantwell, J. and P. Barrera (1995), Intercompany agreements for technology development; lessons from international cartels, *International Studies of Management and*

Organization 25, 1–2, 75–95

Chesnay, F. (1996), Technological agreements, networks and selected issues in economic theory, in R. Coombs et al., *Technological Collaboration*, 18–33

Coombs, R., A. Richards, P.P. Saviotti and V. Walsh (eds) (1996), *Technological Collaboration: The Dynamics of Cooperation in Industrial Innovation*, Cheltenham, Edward Elgar

Hagedoorn, J. and J. Schakenraad (1990a), Strategic partnering and technological cooperation, in B. Dankbaar et al. (eds), *Perspectives in Industrial Organization*, 171–91, Dordrecht, Kluwer

Hagedoorn, J. and J. Schakenraad (1990b), Leading companies and networks of strategic alliances in information technology, *Research Policy* 22, 163–96

Hagedoorn, J. (1993), Understanding strategic technological partnering, *Strategic Management Journal* 14, 371–85

Haklisch, C.S. (1986), Technical alliances in the semiconductor industry, mimeo, New York University, New York

Hippel, Eric von (1988), *The Sources of Innovation*, New York, Oxford University Press

Kleinknecht, A. (1996), Innovation, imitation and R&D cooperation: the Netherlands compared to five other countries (in Dutch), in F.J.M. Zwetsloot (ed.), *De Markt voor Wetenschappelijk Onderzoek*, Utrecht, Lemma, 209–27

Lundvall, B.-Å. (1992), *National Systems of Innovation – Toward a Theory of Innovation and Interactive Learning*, London, Pinter

Mowery, D. (1989), Collaborative ventures between US and foreign manufacturing firms, *Research Policy* 18, 19–32

Pavitt, K. (1984), Sectoral patterns of technical change: towards a taxonomy and a theory, *Research Policy* 13, 343–73

Teece, D.J. (1986), Profiting from technological innovation, *Research Policy* 15, 6, 78–98

Teece, D.J. (1992), Competition, cooperation, and innovation; organizational arrangements for regimes of rapid technological progress, *Journal of Economic Behavior and Organization* 18, 1-25

Wolek, Francis W. (1979), Co-operative research and development in the United States, in: M.J. Baker (ed.), *Industrial Innovation; Technology, Policy, Diffusion*, Macmillan, London, 151–61

9. The organization of vertical technological cooperation

1 INTRODUCTION

As already announced in Chapter 8 this chapter is concerned with specific forms of technological cooperation between manufacturer and supplier in more detail. As discussed in the previous chapter, these forms of cooperation are called 'vertical cooperation'. The purpose of this chapter is to understand why a specific partner is chosen, and how specific organizational designs relate a number of factors, such as the technology, the market, and the goals one or all partners have with the cooperation, and other characteristics of the firms involved.

The empirical part describes specific cases, and how, for example, the (large) manufacturer supports technological learning of the supplier. In the theoretical section we extend the transaction cost analysis of Chapters 3 and 8 by giving a theoretical explanation for vertical technological cooperation.

2 THE EMPIRICAL SIDE

In this section we discuss different forms of vertical cooperation in more detail than in the previous chapter. We need to find a delicate balance between specific features of various cases and the generalizability of the findings.

The increase in outsourcing

In large databases on technological cooperation no systematic distinction is made between vertical and horizontal cooperation. From Chapter 8 double sourcing is the only form that can directly be linked to vertical cooperation. Joint ventures and technology agreements were presented as the most prominent forms of cooperation. Joint ventures are usually not between firms with a vertical relationship, but technology agreements could be between a customer and some of its most advanced suppliers. Technological cooperation with suppliers is usually limited to suppliers that make non-standard goods. The purchase of non-standard goods has been rising particularly significantly in the last decades. Suppliers who deliver non-standard goods are called 'subcontractors'.

One of the main reasons for the increasing significance of subcontracting is the trend towards 'back to the core business'. With the accelerating pace of technological development in many markets and the increase in the number of technologies 'combined' to produce new products, strongly diversified firms are confronted with rising problems of keeping up with all the technologies relevant for their markets. As a result many large manufacturing firms out-source parts of their original production, including the production of parts and subsystems of final products. By implication manufacturers buy more prod-ucts and services from subcontractors.

There are different types of subcontractor, and not all are equally impor-tant for technological cooperation.

Different subcontractors

A difference can be made between a 'jobber', a 'co-developer', and a 'main sup-plier'. A jobber is a supplier who, on demand of the manufacturer, makes a specific product. This product is completely developed by the manufacturer, including test production. This manufacturer defines technical specifications for the product and for the production process on the base of which the job-ber makes a certain volume of the product. One could say that the manufac-turer hires production capacity from the jobber.[1]

A co-developer is a supplier who produces a product on full technical specification from the manufacturer, but it may to a certain extent 'solve' questions of production, logistics, and so on. If price, quality, and delivery scheme of the product are according to what the manufacturer wants, he may refrain from monitoring the co-developer's production process. It is not unu-sual that the co-developer proposes a change in the product design in order to fulfill the manufacturer's wishes in terms of unit price, quality and the like.

A main supplier is a supplier who is able to come up with a product and a production process on the base of functional specifications of the manufac-turer. In contrast to a co-developer, where the manufacturer completely speci-fies the technical features of the product, the main supplier is only given the functions the product must be able to perform. The main supplier himself is offering a technical solution.

From the perspective of technological cooperation this means the follow-ing. In the case of a jobber technology transfer takes place from manufacturer to the supplier. The jobber therefore learns from the manufacturer, but the manufacturer merely hires production capacity. In the case of a co-developer the supplier learns about technical specifications of the manufacturer's prod-uct and is subsequently 'forced' to come up with technological solutions regarding the production process of this specified product. As said above the manufacturer may sometimes learn about alternative product designs from the co-developer. In this respect mutual technological learning is possible, al-though normally the supplier will learn more from the manufacturer than the other way around. In the case of a main supplier a more equal level of technological capabilities exists. As the required product from the supplier is usually a subsystem or system used in the final product of the manufacturer, the technical solutions of the main supplier may give rise to adjustments in the original product design of the manufacturer. Because of the interdependence between final product and (sub)systems, main suppliers are quite often cooperating with the manufacturer in the early stages of the latter's product development process.

Examples of technological cooperation with suppliers[2]

We discuss two examples of cooperation between manufacturer and supplier(s) that shed some light on the specific nature of this form of cooperation and the criteria with which manufacturers choose partners from their often large group of suppliers.

1. European airframe manufacturers and their main-suppliers

The first case is from Paliwoda and Bonaccorsi (1994). It concerns the relationships between airframe manufacturers and their suppliers.

The design process in the airframe industry is characterized by many 'unknowns' (ibid., 235). To avoid costly negotiations with suppliers the industry used cost-plus contracts to enable suppliers to fund unexpected development and design activities. There is, however, a trend towards fixed-price contracts, that force suppliers to bear all the financial risks of development. This will be aggrevated when suppliers tend to quote a price that does not cover development costs in order to win the contract. According to transaction cost theory, which we discuss in the next section, the chance that suppliers forced into this position will behave opportunistically is paramount. In this industry it would mean that when the manufacturer changes its original design (and thereby imposes new demands on the design of subsystems of suppliers) suppliers will take the opportunity to claim excessive additional development funds. The more the airframe manufacturer is unfamiliar with the supplier's technology or when the technology as such bears great uncertainties, the less he can judge whether the supplier's claim is fair.

Although the structure in the industry and its specific contractual relations may favor a form of opportunism, in practice the situation as described (ibid. 236) is 'mitigated':

> the relationships between suppliers and customers are based on a high degree of mutual trust. Information exchange is very great, so customers can more easily detect opportunistic behavior; they think they can 'sense' when suppliers are cheating. They also know that sometimes drawing modifications are requested by suppliers,

so mutual compliance should be expected. They often have a continuous relationship with suppliers, so the possibility of detection and loss of the next order is deterrent.'

There is a clear tendency towards single sourcing and a reduction of the number of suppliers. Power systems and avionics (for example engine and wing respectively) are generally single sourced. They are already quite complex systems on their own. Systems bought from a number of 'preferred' suppliers are now common in the European airframe industry. This happened because some suppliers choose such a strategy explicitly, but also because of 'product reliability'. 'Liability claims are the main area of conflict between airframe manufacturers and suppliers. Systems buying is a means of reducing this problem, because by definition, a system supplier is responsible for the integration and performance of a set of different components' (ibid., 237). So, despite the fact that the airframe manufacturer is unable to estimate the development costs of the components and the integration costs of the system, systems buying is preferred to single-source buying of all components and assembly in-house to avoid product liability problems. Multiple sourcing is only undertaken in the area of 'non-system-related' components and even there the number of suppliers has been reduced. Apparently multiple sourcing of system-related components and subsystems by the airframe manufacturer (and subsequent assembly in-house) is not an alternative.

Approximately 80 percent of the total purchases fall within the long-term business category. These long-term agreements with single source system suppliers cover more than the batches of production in former contracts. In addition supplier involvement is very strong in addressing the earlier mentioned uncertainties in product development at both sides. The advantage of a single source is also that the airframe manufacturer can learn effectively about the integration tasks of the supplier; the fixed-price contracts may therefore be a good estimate of real costs. The lack of competition in the supply market is usually of great concern for cost-reduction strategies that are also adopted in this industry. A surprising aspect of customer–subcontractor relationships here is that:

> At the moment, customers are reluctant to award a share of business to non-established suppliers; they prefer to work with preferred vendors in order to obtain cost reductions from them. Their approach would be, 'how can you, our preferred supplier, and we, the customer, work together to obtain the price offered by your competitor?'
>
> The core of the issue here is the high level of interaction required between the design process of the customer and its important suppliers. Airframe companies normally involve suppliers within one year after the launch of a new aircraft

program and quite often even before the public announcement is given. The Airbus A330 and A340 had their first flights in 1992, but supplier involvement began in 1986. (ibid., 239)

In this industry it must be realized that even for main suppliers the functional requirements are set by the customer. 'The definition of requirements is, in itself, the result of continuous interaction with suppliers. Technical discussions with several suppliers are opened long before specifications are written and the request for proposals is issued . . . During this period, all suppliers involved invest their own money in preliminary design activity and in solving the technical problems of the customer. (ibid., 239)

2. The strategy of a high-tech supplier towards machine manufacturers in the aircraft industry

In the previous case it became clear that engine manufacturers are 'single-source-preferred-suppliers' to the airframe manufacturers. The focus on the relationship was from the point of view of the customer. The engine manufacturers and other main suppliers have their own network of suppliers, hence they themselves purchase custom-made products. Frear and Metcalf (1995) examine the position and strategy of a cast-products supplier of high international reputation to the aircraft engine manufacturers. This case is interesting to compare with the previous one in a number of respects. First, the cast producer supplies to some of the main suppliers of the airframe manufacturers and is thus part of their supply chain. Second, the focus is on the supplier snot on the customer. More generally it is interesting from a scientific point of view to examine both sides of a business to business relationship. Third, the subcontractor–customer relationship is analyzed from the perspective of a larger network in which it is embedded and it is shown how this network influences the subcontractor–customer relationship.

The 'focal relationships' of the case are of Cast Products (CP), a manufacturer of complex steel and aluminium castings, and two main suppliers of engine manufacturers of the airframe industry, FiatAvio (FA) and Hispano-Suiza (HP). According to the authors the whole civilian airplane 'cluster' is of importance to understand a particular kind of relationship within the cluster. Figure 9.1 gives a simplified picture of this cluster. It shows types of players with some examples of companies, including the three mentioned.

The situation in the cluster in the early 1990s can briefly be described as follows. The airline companies have postponed or even cancelled a considerable number of orders. As a consequence demand for the airframe manufacturers has been low and uncertain. 'These reductions cascade down through the entire manufacturing chain. Every aircraft not produced means that, at least, two fewer aircraft engines plus spare parts are not manufactured. All

Figure 9.1 The main types of players in the civilian airplane cluster and the position of the case actors

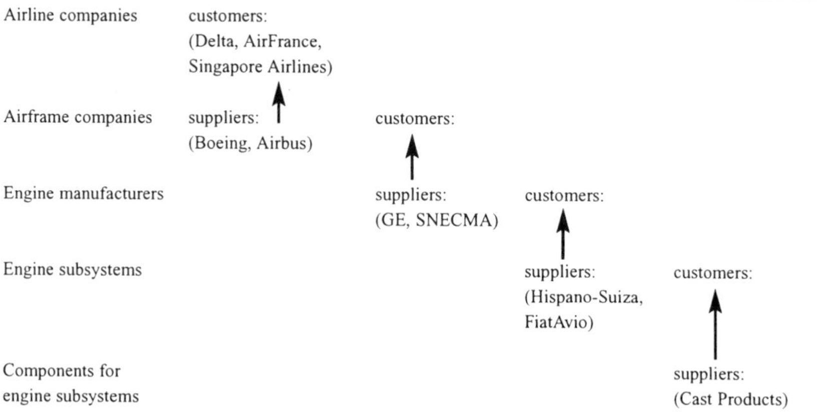

Airline companies	customers: (Delta, AirFrance, Singapore Airlines)			
Airframe companies	suppliers: ↑ (Boeing, Airbus)	customers: ↑		
Engine manufacturers		suppliers: (GE, SNECMA)	customers: ↑	
Engine subsystems			suppliers: (Hispano-Suiza, FiatAvio)	customers: ↑
Components for engine subsystems				suppliers: (Cast Products)

Source: based on figures in Frear and Metcalf (1995)

four major aircraft engine manufacturers (GE, Pratt-Whitney, SNECMA, and Rolls Royce) have undergone major restructuring to enhance efficiencies and reduce costs' (ibid., 381). Normally this would have had even more severe repercussions for suppliers lower in the hierarchy, such as FA and HP, and further down CP. However, these second and third tier suppliers are not typical; FA is part of the FIAT Group and HP is a subsidiary of SNECMA. The cluster is even more complex, because a relatively large number of strategic alliances and collaborations exist. Several examples are mentioned in the case: Airbus (Aerospatial-It, Deutsche Aerospace, British Aerospace, and CASA -Sp), the cooperation between Boeing and Deutsche Aerospace in the Jumbo Jet, MBB and Fokker (no longer in existence), in the aircraft engine segment: GE/SNECMA (as CFM International), GE/SNECMA/IHI/FA for the development of the GE90 engine, and BMW/Rolls Royce for the BR 710.

CP is a business unit of Teledyne Aircraft Products, an operating company of Teledyne Inc., a diversified manufacturing company with 21,000 workers and sales in 1993 of 2.4 billion dollars. For the purpose of this chapter it suffices to explain the relation between CP and HP in the context of the larger network(s).

Before the subcontractor–customer relationship is further examined, another main characteristic of the airframe industry needs to be emphasized: its extremely high level of product development costs. The development costs for the new jumbo jet are estimated at over 15 billion US dollars. The combination of low market demand and high development costs in this industry is a classical example of low market incentives and small market room to innovation (Scherer, 1984). Due to the relatively small number of players with a sufficient technological expertise all kinds of alliance and even partly overlapping networks may arise.

HP is a subsidiary of Societé National d'Etude et de Construction de Moteur d'Aviation (Groupe SNECMA), the largest manufacturer of aircraft engines in Europe, located in Paris; its sales in 1992 were 4.3 billion dollars and it employed 24,911 workers. HP, also located in France, manufactures gearboxes for aircraft engines (among others). The relationship with CP began in 1991 through an older relationship of ALLVAC (part of Teledyne Inc.) with CP. In addition the purchasing manager of HP knew CP through his previous position in the parent company SNECMA. CP supplies HS with the gearbox case casting that houses the components and lubricant for the aircraft gearbox. HS machines the castings, assembles the components, adds the lubricant, and ships the completed gearbox to CFMI for installation into the aircraft engine. The casting process for this application is complex and only five to six foundries in the world have the know-how to perform it. HS sets the technical specifications of the casting, but CP is completely responsible for the process to make it (which makes CP a high-value co-maker, but not a main supplier). In addition to the physical product CP provides HS with technical assistance. 'CP has committed considerable resources to its relationship with HS. In fact, the entire first year's contract was devoted to building tools [partly because European and US design standards differ]' (ibid., 384). CP sees HP as a very important customer, because through her it may get access to SNECMA's two networks in the aircraft engine market, which together have 50 percent market share. HS also devotes sizeable resources to the relationship; transaction costs are high because it re-certifies and qualifies CP as a supplier of each new product cycle, which normally is annually. 'Entering into a long-term agreement with CP would not only enable HS to reduce these transaction costs but also further cement the relationship. CP is pursuing this path' (ibid., 384). There is a high degree of cooperation by both parties; technical and pricing information are exchanged frequently. There have been language and time zone problems in communication and there have been some problems in the start-up period, especially about tooling specifications and quality requirements. In general, however, 'the relationship has matured into a very positive one . . . [and] CP has complete confidence in the technical advice and information provided by HS . . . CP believes that HS is a completely trustworthy partner that will honor the commitments that it makes' (ibid., 385). The operational connections are 'supported' by extensive interaction between high-level managers at both sides. Engineers from CP visit HS regularly.

3 THE THEORETICAL SIDE

What kinds of economic explanation can be given for the various forms of vertical technological cooperation? Transaction cost economics (TCE) gives an explanation for cooperation more generally, as has been shown earlier in this book. Depending on the specific circumstances a firm will coordinate transactions internally, will purchase the goods on the market, or will choose an intermediate form of coordination of the transaction, that is, cooperation. It has been argued that transaction cost economics cannot explain why sometimes a joint venture is chosen, and sometimes a technological agreement or another form of technological cooperation.

Vertical cooperation according to TCE

In the previous chapter we already discussed how a manufacturing firm might learn from its suppliers and vice versa. In the empirical section of the current chapter it was stated that the different types of supplier have different kinds of technological cooperation with the manufacturer.

Although TCE focuses on minimization of transaction costs, it is also used to explain at least some aspects of technological cooperation between manufacturer and supplier (see the quote of Hagedoorn, 1993, on TCE in Chapter 8).

In a 'pure' transaction cost reasoning technological cooperation between customer and supplier might be perceived in two alternative ways.[3] First, is it included in the more general coordination and monitoring of tasks, required when a manufacturer sources out some activity for which the supplier has to make specific investments; in other words, technological cooperation is included in the 'governance' the customer undertakes with respect to a specific supplier in order to minimize transaction and production costs. Second, is to see technological cooperation with a supplier as a conscious 'buy-decision' of the customer with respect to R&D activities. The make or buy decision has two steps in this case. Step one is the decision to develop a specific product that is a part or subsystem of the final product of the manufacturer in-house, or to find a supplier who can develop it. Step two concerns the decision who produces this part/subsystem, the manufacturer or a supplier. In the first explanation technological cooperation remains implicit. In the second explanation technological cooperation follows from the decision to source out the R&D activity for the development of part of the product. In my opinion this more explicit explanation is more attractive, also because it seems to relate neatly to some of the cases of subcontracting in practice. Two points need clarification, however. First, as we have seen, different types of supplier exist and the question is whether TCE can differentiate between them. Second, what exactly does 'technological cooperation' mean in TCE terms? In the above it was suggested that the manufacturer could decide to find a supplier for the 'production' of a specific part of its R&D. But cooperation in TCE implies more than contracting a supplier. We begin with this last point.

As has been explained before, cooperation is assumed to be cost-efficient when asset specificity, frequency, and uncertainty have values 'between high and low'. In the context of innovation, a transaction may imply two things: first, the delivery to the manufacturer of the 'final product' of the innovation process, that is the new part developed by the supplier; second, the exchange of information between manufacturer and supplier needed to let the supplier develop the new part. In the first case the innovation process is 'reduced' to a regular delivery of a product from a supplier to the manufacturer, hence the normal transaction cost considerations apply. In my opinion the second interpretation is therefore more fruitful.[4]

The question remains what medium values of frequency of information exchange, asset specificy of this information, and uncertainty with respect to the behavior of the supplier mean. We come back to that when discussing the various types of supplier.

Monteverde and Teece (1982a and 1982b) argue that when information about the manufacturer's final product and/or its production process needs to be transferred to the subcontractor in order to let the latter develop a part or subsystem most effectively, this supplier might behave opportunistically. For example, the supplier could use the information to make the final product itself. It would thus appropriate part of the innovation profits (as Teece, 1986, argues, this is one of the situations of a weak control of complementary technology). It is therefore clear what the transaction cost problem of technological cooperation with suppliers looks like. Williamson (1985) stresses that the danger of opportunistic behavior is especially severe when asset specificity is high. One might expect that in such cases the manufacturer integrates the supplier and that the three types of subcontractor discussed in the empirical part of the chapter are cases of low or medium asset specificity.

The most common case of a buy-decision concerning a situation of asset specificity is when a manufacturer develops a product and looks for a supplier who can make this product on full technical specification of the manufacturer. Such a supplier has been called a 'jobber'. When both product development and production are sourced out, one speaks of a 'main supplier'. The co-developer is in between. Table 9.1 summarizes these findings.

A manufacturer would, according to the 'logic' of TCE, outsource R&D concerning parts and subsystems of its product, when the sum of production costs (in this case R&D) and transaction costs is lower than in the situation where he would develop it in-house. This could be the case when a supplier has better technological expertise about component technology and the related production line and when the manufacturer estimates that the costs of moni-

Table 9.1 Outsourcing of R&D to the supplier as technological cooperation

Production done by:	Product development done by:	
	Manufacturer	Supplier
Manufacturer	Complete make decision	Rare case
Supplier	Jobber [1] Co-developer [2] (occasionally techn. coop)	Main supplier techn. coop.

[1] Specifications of production process done by manufacturer.
[2] Specifications developed by supplier.

toring this R&D activity of the supplier are relatively low. From this perspective a jobber is no form of outsourcing of R&D. In the case of a co-developer the manufacturer develops the product, but the R&D of the production line for this product is sourced out. In most cases the suppliers can produce the product with minor 'retooling' on existing product lines, so asset specificity is low, or at best medium. The need for information exchange ('frequency') is low, or medium (when the supplier makes suggestions for improving the manufacturers product or process). Because the process technology used by the co-maker is rather standard, uncertainty of its behavior may be expected to be low too.

The case of the main supplier is different. Here the supplier develops a product specifically for the manufacturer. Depending on whether the supplier has been confronted with similar functional specifications to this manufacturer or other manufacturers and whether specific investments must be made in the production line, asset specificity of the R&D can be estimated as medium to high. So-called early supplier involvement is quite common with main suppliers, which would mean medium to high frequency of information exchange. Whether uncertainty for the manufacturer with regard to the behavior of the supplier is high, depends on his experience with and knowledge of the product and process technologies of the supplier. From this reasoning it can be derived that in the case of the main supplier asset specificity and uncertainty are positively correlated: the more specific the knowledge needed to make the (sub)system for the manufacturer (high asset specificity) the less likely it is that the manufacturer can effectively monitor and evaluate the supplier's R&D process (high uncertainty). TCE would therefore predict that main-suppliers only develop systems with medium asset specificity and uncertainty and that systems with high asset specificity and uncertainty are developed in-house. To the extent that the trend towards 'back to the core business' implies outsourcing of product development regarding technological fields which are increasingly less familiar to the manufacturer, the existence of (some of) the main-suppliers can be perceived as in contrast with TCE.[5]

Trust

Private firms try to commercially apply new technology to make profits. While they perceive other firms doing the same, there is generally considerable distrust about cooperation. The more technological cooperation is related to the core of the business the more distrust there will be. Especially in small and medium-sized firms the core business is normally more involved than in larger firms. In addition the owner/director of the smaller firm usually deals with technological cooperation directly and he perceives the 'technological knowledge base' of the firm as his own. In contrast negotiations about technological cooperation in larger firms are often dealt with by professional managers or researchers who identify themselves less with the technological knowledge base of the company. This may explain the importance of trust, especially in cooperation of SMF's, in practice. From the two examples in the empirical section it became clear that long-term relationships with suppliers are preferred by manufacturers which could mean that trust plays an important role.

In the discussions around transaction cost economics (TCE) trust has become a major issue. Williamson argues that trust as it can be found in personal relations is seldom present in business relations, while 'calculated trust' may be seen quite commonly in business transactions (Williamson, 1993). Calculated trust is described as the way in which trusting the other party is assumed to lead to an economic advantage for the 'trustor'.

In terms of calculated trust TCE argues that the manufacturer will use safeguards to avoid opportunism by the supplier. Examples are the provision of technical information by co-location of the supplier's engineers at the manufacturer's development unit, pre-financing of (part of) the supplier's development costs or the promise of future contracts when the current delivery is up to the expected level. The disadvantage of TCE is that it does not make clear to which point safeguards can compensate for increasing levels of asset specificity and uncertainty. It merely predicts that when these levels are high integration of the supply unit will occur.

It could be argued that TCE overemphasizes the concern for opportunism and underestimates the 'dynamics of asset specificity'. With the latter we mean that specialization of assets is gradually taking place on both sides of the partnership. It needs a certain intensity of the relationship to build up asset specificity. The danger of opportunism is one side of the coin, being able to communicate technological issues and to learn from each other is the other side. As a consequence the level of trust, or put more mildly the institutional setting between manufacturer and supplier, has been developed in such a way that 'integration' is an unnecessary costly decision. In addition integration forecloses the opportunity to select another partner with 'better' technological expertise in the near future. The case below discussed in terms of the capabilities approach provides more insight into the limitations of TCE.

Capabilities approach and subcontracting

Argyres (1996) discusses two products made by a large corporation for which some important make or buy decisions are contemplated by management. These make or buy decisions concern various activities: the design of the products, the manufacturing of a number of components, the assembly process itself, and the making of molds for one of the product's manufacturing processes.

The data on the various activities are used as qualitative case study material to examine whether TCE and/or the 'capability approach' to the firm can explain why in one case the activity is sourced out and in the other carried out in-house.

The corporation operates in many markets, that can all be characterized by technology competition; a firm is normally offering a different technological solution than its competitors in the market. Hence, the firm under view operates on the basis of specific technological capabilities, which in this case are material-science based.

Although all the cases discussed by Argyres are instructive to comprehend the differences between the ability and the TCE approach, only the ones embodying elements of technological cooperation are discussed here.

> The capabilities approach to the firm postulates that firms vertically integrate activities for which they possess capabilities that are superior to potential suppliers . . . The claim is that because firms have different capabilities, they often carry out the same activity with different production costs. Economizing firms will take this into account when deciding whether to perform the activity in-house or on a contract basis with another firm. As relative capabilities change, firm boundaries are adjusted accordingly. (Argyres, 1996, 129)[6]

Capabilities are based on specific technological and managerial knowledge. It takes time for firms to acquire the capabilities to execute a specific activity.

To fully appreciate the differences with TCE it is useful to discuss the role of production costs, following Argyres (ibid., 130). Decision makers in TCE minimize the sum of production and transaction costs. Transaction costs with regard to internal suppliers are normally lower than with regard to outside suppliers. Production costs of the internal unit may be higher than of the outside supplier because the latter can gain more from economies of scale. Internal units, however, can also sell to competitors and the only thing to prevent that from happening is management's fear of opportunism by these competitors. As a result, one can conclude that transaction costs considerations are mainly influencing make or buy decisions in TCE. The difference with the capabilities approach becomes clear (ibid. 130):

The capabilities approach contradicts transaction cost theory by suggesting that differential production costs play an important, independent role in make-or-buy decisions, and arise from different firm-specific capabilities rather than from scale economies.

Firm-specificity follows from the exchange of information between individuals through which the capabilities are gradually developed; accumulation of technological knowledge and the build-up of procedures and managerial skills to exploit this knowledge play an important role. One could speak of the 'co-evolution' of technology and institutions within the firm. These capabilities are difficult to imitate and therefore they constitute a precisely competitive advantage for the firm. An example of such a distinct capability could be a firm's accumulation of technological and managerial knowledge as reflected in particular product development projects (such as the biotechnology research program of a pharmaceutical company).

To shed more light on the role of capabilities in production casts, Argyres studies the production chain of two relatively simple products: the production of a cable connector and the production of specialty wire and cable. We discuss the design of the CC2 and the mold-making for the 'precision injection molding' process with which two of the three components of the connector are made. The 'CC2' cable connector is one of a set of cable connectors that were recently introduced by the firm. The firm (which is called 'TightFit' by Argyres) did not have previous experience with the technologies related to this product. Of importance to this chapter is the following description of the development of this product: 'The product design process was carried out entirely inside TightFit's R&D labs, while product testing and design evaluation occurred in the Telecommunications Division lab by technicians newly transferred from industrial tubing development' (ibid., 132). The firm tried to develop a new connector based on proprietary materials science technology, but it failed. After the failure of the first design engineers worked for another two years on a second design based on other technologies. Although the product's technology was outside the firm's core competencies, it was expected to give access to customers' needs on technological issues that could become closely related to current capabilities. What remains, however, is Argyres' assessment (ibid., 134) that 'no previously accumulated firm-specific knowledge was used in the CC2 which was reflected in the form of proprietary materials science technology'. At the same time the knowledge gained was specific enough to acquire a number of patents.

The mold-making process is quite in contrast with the design process of the CC2. The former is carried out by an external supplier who makes the molds for several TightFit divisions. The plastic injection molds are custom designed and built for every part in the customer's product(s).

> 'Mold-making for precision plastic injection involves design of the metal mold and then building it using special craftsman's tools (which are not specific to particular designs). These are very specialized skills, requiring as many as 7 years apprenticeship under an experienced mold-maker. Very little formal schooling is available, or necessary, for training in mold-making'. (ibid., 135)

All but one mold-maker in the supplying firm have at least 14 years of experience and working together in the process. This team-working is often needed, because an initially designed mold normally does not produce a product which matches the drawings of the customer, despite the vast experience of the mold-maker. Colleagues may help in subsequent 'trials'.

Before we discuss the main empirical results of the study the method used will be discussed briefly. The data were gathered in interviews with managers, engineers, and chemical specialists from various divisions of the firm. In addition company documents were studied. The products to be studied were not selected at random, but followed from the firm's involvement in make-or-buy decisions during the investigation. When it turned out that capabilities were playing a role in the organization of successive stages of the production processes studied, more information about them was gathered. How the interviewees were selected and how the interviews took place is not described.

After the failure of the first design for the connector one shifted to the use of technologies that were outside the firm's core competencies, although 'this decision to design, test and evaluate in house does not appear to have benefited from a deliberate decision-making process' (ibid., 132).

To develop the connector TightFit could have decided to use an outside supplier who did have core competencies in the field. However, a large portion of the development time (2 years) 'once spent, could not have been easily transferred to alternative uses, if, say, it was spent by a supplier dealing with a buyer who was engaging in hold-up. Thus, one can interpret the decision to design and develop the product in-house as associated with the high level of human asset specificity necessary to the activity' (ibid., 134). According to TCE a supplier anticipates such a situation by asking which safeguards, may be the reason for the customer to start development in-house. Argyres concludes that the existing capabilities of TightFit cannot explain this make-or-buy decision. In this case the decision to enter the market for cable connectors made it necessary to build up the new capabilities in-house, while outsourcing would result in very high transaction costs because monitoring the supplier could not be effectively done on the basis of existing capabilities. In contrast, the mold-making process provides an example of high asset specificity for which TCE predicts internalization. Following Argyres' arguments (ibid., 138), the capabilities-approach stresses the different nature of technological and managerial competencies in innovative design activities, which are the core of TightFit's activities, and in mold-making.

For example, innovative design activities and mold making activities proceed with different underlying models of how to develop new products. The first uses functional novelty as the primary criterion to judge alternative designs, while the latter uses manufacturing ease (cost). In addition, innovative design relies on academic training in principles of physics, chemistry, engineering, and experimental techniques. Mold-making, on the other hand, relies on rules of thumb and notions not firmly based on abstract knowledge. The tacit component is more important than in connector design.

Because the technological and managerial capabilities of mold-making and TightFit's core competencies were so different in-house production was not considered, also because training people in the process of mold-making requires many years. Acquiring such a supplier would have been an alternative, looking at TightFit's demand for molds, but significant time would be necessary to absorb the acquisition.

Making molds is a peculiar example of the customer setting functional, not technical, specifications. On the one hand the customer does specify the technical specifications in the drawings provided to the supplier, but these are merely the base for a trial-and-error process of making molds which are able to produce the products in conformity with the technical specifications. So we are talking about a part used in the production process of the cable connector for which no technical specifications are set by the customer. Because of high asset specificity TCE predicts in-house production, while the capabilities approach argues that the capabilities required for the making of molds is too different from the current expertise of TightFit.

A main feature of the mold-making process is that it does not require a considerable amount of communication on technological issues between TightFit's engineers and R&D personnel and the mold-makers. In contrast modern product development activities are characterized by extensive interaction between people with different functional backgrounds, when products are complex. The capabilities approach, as presented by Argyres (1996), makes clear that, to the extent that product development activities fit well to the core competencies, they will be undertaken in-house. The example of the making of molds for the manufacturing process of the cable connector by an outside supplier cannot be explained by TCE. The capabilities approach opens up a new lane to understand more fully the existence of main suppliers with whom customers communicate extensively, but who are not internalized.

In many markets technological specialization is augmenting and the numbers of different technologies, which are combined into new products, increases. The capabilities approach offers an explanation why firms concentrate on a limited number of technological capabilities and why it is so difficult to acquire new ones, even when the market incentives to do so are very great. Under these conditions, firms are forced to interact, extensively with some of their suppliers. The approach thus provides some theoretical explanation for the 'technological interactions' described in the case studies on the aircraft industry. Similar interactions are discussed in terms of 'asset co-specialization' in TCE-related studies (see, for example, Dyer, 1996). The concept of abilities does not consist of a set of well-defined behavioral assumptions and environmental conditions, comparable with TCE. Therefore it is difficult to connect, at a theoretical level, asset co-specialization in a supplier's network with the combination of capabilities of various autonomous firms, although, clearly, similar issues are discussed.

4 CONCLUSIONS

Vertical technological cooperation concerns a manufacturer and some of its customers and suppliers. The studies of von Hippel emphasize the role of the customer; here the focus has been on suppliers. In particular main suppliers have cooperative agreements with manufacturers in which both parties learn technological things from the other. From the several cases discussed in this chapter it became clear that long-term relationships exist between the manufacturer and the supplier. Although one should not disregard 'a business-like matter' trust between the manufacturer (customer) and supplier appears to be important. Transaction cost theory emphasizes opportunism and the need for monitoring of the suppliers with which one cooperates. The weakness of this approach is that it predicts full integration of suppliers when asset specificity is high. This contradicts practice where cooperation with 'advanced' suppliers involves very specific technologies. The abilities approach argues that the manufacturer often has difficulty in developing new abilities and that therefore cooperation with specialized suppliers might be advantageous.

NOTES

1. Quite often the manufacturer produces the product itself and the jobber's capacity is hired when its own capacity is fully utilized.
2. The examples given here are from Beije (1998)
3. Williamson (1985) clearly separates transaction cost motivated organizational solutions from innovation. He states that a tradeoff must be made between governance structures which minimize production plus transaction costs and the development of new products.
4. That it is indeed an interpretation of TCE follows from the fact that TCE, as developed by Williamson, is focusing on transactions concerning an existing product, and not on the development of a new product, as Williamson himself explicitly states.
5. A dynamic perspective, which is lacking in TCE, could provide additional understanding; in several cases the author is familiar with, large manufacturers select main suppliers on the base of the knowledge and experience present from the time when they still developed and manufactured the product that was outside the 'core business'. Once the main supplier's technological expertise is slowly 'outgrowing' that of the manufacturer a trust relationship has to be built up.
6. Argyres refers to Demsetz, 1988; Teece, 1988; Kogut and Zander, 1992; and Langlois, 1992

REFERENCES

Argyres, N. (1996), Evidence on the role of firm capabilities in vertical integration decisions, *Strategic Management Journal* 17, 129–50

Beije, Paul R. (1998), Technological cooperation between customer and subcontractors, *Handbook of Industrial Marketing, Advances in Business Marketing and Purchasing* vol. 8, Greenwich (CT), JAI Press (forthcoming)

Demsetz H. (1988), The theory of the firm revisited, *Journal of Law, Economics and Organization* 1, 141–61

Dyer, Jeffrey (1996), Specialized supplier networks as a source of competitors advantage: evidence from the auto industry, *Strategic Management Journal* 17, 271–91

Frear, Carl R. and Lynn E. Metcalf (1995), Strategic alliances and technology networks: a study of a cast-products supplier in the aircraft industry, *Industrial Marketing Journal* 14, 5, 379–90

Hagedoorn, J. (1993), Understanding the rationale of strategic technology partnering: interorganizational modes of cooperation and sectoral differences, *Strategic Management Journal* 14, 371–85

Kogut, B. and U. Zander (1992), Knowledge of the firm, combinative capabilities, and the replication of technology, *Organization Science* 3, 383–97

Langlois, R. (1992), Transaction cost economics in real time, *Industrial and Corporate Change* 1, 99–127

Monteverde, K. and David J. Teece (1982a), Appropriable rents and quasi-vertical integration, *The Journal of Law and Economics* 25, 321–28

Monteverde, K. and David J. Teece (1982b), Supplier switching costs and vertical integration in the automobile industry, *Bell Journal of Economics* 12, 206–13

Scherer, F.M. (1984), *Innovation and Growth: Schumpeterian Perspectives*, Cambridge (MA), MIT Press

Paliwoda, Stanley J. and Andrea J. Bonaccorsi (1994), Trends in procurement strategies within the European aircraft industry, *Industrial Marketing Management* 23, 235–44

Teece, David J. (1986), Profiting from technological innovation: implications for integration, collaboration, licensing and public policy, *Research Policy* 15, 6, 285–305

Teece, David J. (1988), Technological change and the nature of the firm, in: G. Dosi, C. Freeman, R. Nelson, G. Silverberg and L. Soete (eds), *Technical Change and Economic Theory*, Pinter, London, 256–81

Williamson, Oliver E. (1985), *The Economic Institutions of Capitalism*, New York, The Free Press.

Williamson, Oliver E. (1993), Calculativeness, trust, and economic organization, *Journal of Law and Economics*, 36, 453–86

10. The organization of horizontal cooperation

1 INTRODUCTION

Cooperation between firms in the same industry may seem curious because these firms are, at least potentially, direct competitors. A large part of the number of technological cooperations discussed in Chapter 8 were 'horizontal' ('between competitors') cooperations. One of the earlier empirical accounts of cooperation (Mariti and Smiley, 1983) gave the same picture, although intra-industry cooperative agreements dominate in some industries (electronics, telecommunications, automobile) with a high frequency of cooperative agreements, while in most industries this frequency seems to be much lower, and in some of the latter inter-industry agreements are the majority (although not frequent). The empirical part describes specific cases and how, for examples the sharing of costs and benefits is undertaken. The theoretical part examines economic theories related to 'horizontal' cooperation, which address the organization and other details presented in the empirical part.

2 EMPIRICAL FINDINGS

From the analysis of patterns of technological cooperation has resulted the insight that firms often begin to cooperate with a loose form, that can be terminated without a great loss of resources, and then proceed with a tighter form of cooperation with a partner they, in the meantime, have learned to know. Nevertheless, cooperative agreements not only differ in terms of intensity of investments, but also in terms of the specific technological goals. We have seen that market access and influence on market structure were important motives for technological cooperation. Here we concentrate on the technological side. Our guideline in discussing specific cases is not the underlying technological motives, but the intended technological outcomes. In this respect a difference can be made between cooperations where new products (or processes) are the result of joint efforts, and cooperations where the outcome is some kind of generic technological knowledge that can subsequently be commercially applied. This gives rise to the following dichotomy shown in Table 10.1.

In this table examples of technological cooperation are mentioned. The US minimills case and the MCC case are discussed below. The technology agree-

Table 10.1 Different forms of technological cooperation

Form of cooperation	Application of technological knowledge	
	Commercialization	Generic base
Loose	US minimills	Technology agreements Hoffman-La Roche
Tight	Eureka projects, JV's	MCC corporation

ments of Hoffman-La Roche mean the relationships with several biotechnology firms. Some of these agreements are made in terms of minority stakes. The goal is the strengthening of the firm's generic knowlegde base with respect to biotechnology. Eureka projects will be discussed in the last chapter. The Eureka program is a technology program initiated by the EU in which firms and research institutes from several member states jointly undertake R&D aimed at commercialization of the technologies involved. Some of the projects in this program are tight in the sense that extensive contracts are designed between the partners. Most technology joint vertures are more representative of tight forms of technological cooperation with a commercial goal. In the cases of the airplane market in the previous chapter two examples were mentioned of joint ventures on the development and commercialization of new engines.

The US minimills case

As described by von Hippel (1987) and Schrader (1991) a large number of so-called minimills exchange information about process innovations. Minimills compete with the large steel plants, but with a different technology to make steel and with a different input. While the large steel plants use iron ore and coal as inputs, the minimills convert scrap into steel. The process technology to do so is quite different from the traditional technology of steel-making and although minimills differ considerably in their capacity (tons of steel per year), they are on average much smaller than the original steel manufacturers. There are approximately 60 steel minimill plants (and approximataly 40 producers) in the US and 'the most productive of these have surpassed their Japanese competitors in terms of tons of steel per labor hour input, and are regarded as among the world leaders in this process'. (von Hippel, 1987, 135) They also compete effectively against the major integrated US steel producers, on the basis of modern facilities and relatively low labor, capital, and materials costs.

What kind of knowledge is exchanged, and how does this exchange of technological information takes place? Von Hippel uses the term 'know-how trading'. Know-how is the

> accumulated practical skill or expertise which allows one to do something smoothly and efficiently . . . that [is] held in the minds of the firm's engineers who develop its products and develop and operate its processes. Often, a firm considers a significant proportion of such know-how proprietary and protects it as a trade secret. (von Hippel, 1987, 133–4)

Proprietary means that the information is firm-specific and potentially provides a competitive advantage. When a firm lacks some proprietary information to solve problems in, for example, its process, the engineers cannot find it in publications. The information source the firm can utilize are professional colleagues in firms that make similar products or use similar processes. 'But are such professional colleagues willing to reveal their proprietary know-how to employees of rival firms? Interestingly, it appears that the answer is quite uniformly 'yes' in at least one industry, and quite probably in many'. (ibid., 134)

What about the way the know-how is traded? There is no official agreement, let alone a formal cooperative organization like a joint venture. Engineers involved with continuous improvement in the production process are confronted with specific problems and they regularly have contact by telephone with engineers from other minimills about these problems. Management did not develop a specific strategy to exchange information, engineers simply ask colleagues from another minimill for information about process technology whenever necessary, and they expect to provide information to the other firm on the next occasion. The next quote gives an almost complete picture (ibid., 134):

> The informal know-how trading behavior which I have observed to date appears to involve informal trading 'networks' which develop between engineers having common professional interests. Network formation begins when, at conferences and elsewhere, an engineer makes private judgments as to the areas of expertise and abilities of those he meets, and builds his personal informal list of possibly useful expert contacts. Later, when 'Engineer A' encounters a product or process

development problem he finds difficult, he activates his network by calling 'Engineer B', an appropriately knowledgeable contact who works for a directly competing (or non-competing) firm, for advice. B makes a judgment as to the competitive value of the information A is requesting. If it seems to him vital to his own firm's competitive position, he will not provide it. However, if it seems useful but not crucial - and if A seems to be a potential useful and appropriately knowledgeable expert who may be of future value to B – B will answer his request as well as he can/or refer him to other experts of his acquaintance. B may go to considerable length to help A: He may, for example, run a special simulation for him on the firm's computer system. At the same time, A realizes that in asking for and accepting the help, he is incurring an obligation to provide similar help to B – or to another referred by B – at some future date. No explicit accounting of favors given and received is kept in instances studied to date, but the obligation to return a favor seems strongly felt by recipients.

In the study the four largest minimills and six additional ones, randomly selected from the complete list of minimills, were studied. During some interviews Quanex Corporation was suggested as an interesting outsider, because it was supposed not to 'trade know-how'. Table 10.2 provides information about the development of proprietary know-how in terms of major or minor process developments made in-house and the existence of know-exchange with other minimills.

Table 10.2 Know-how trading patterns

Steel minimill firm	In-house process devel.	Know-how trade?
Four largest firms		
Chaparral	Major	Yes
Florida Steel	Major	Yes
North Star	Major	Yes
Nucor	Major	Yes
Other		
Bayou Steel	Minor	Yes
Cascade Steel	Minor	Yes
Charter Electric	Minor	Yes
Kentucky Electric	Minor	Yes
Marathon Steel	Minor	Yes
Raritan River	Minor	Yes
Quanex	Minor	No

Source: von Hippel (1987)

Figure 10.1 Importance of informal information transfer between firms for respondents

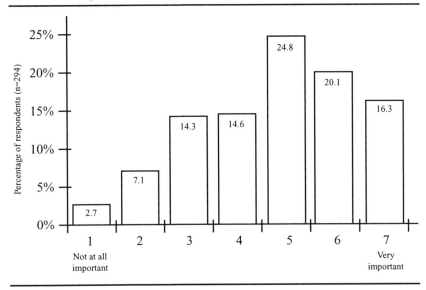

Source: Schrader (1991)

In addition to the information exchange already described know-how trading may also take the form of training (free of charge) of operating employees of competitors, or firm personnel were sent to the plants of competitors to help set up unfamiliar equipment (ibid., 135). It is important that firms report that engineers are selective and that they only exchange information with other minimills when they are assumed to have a similar level of expertise. This must be seen in the light of considerable differences in technical accomplishment among the firms interviewed by von Hippel (ibid., 136).

Schrader (1991) studied a larger set of minimills and included specialty steel firms.[1] A questionnaire was sent to 477 technical managers from 127 firms. The number of responding managers was 294. The findings of von Hippel that know-how trading is important were confirmed. On a 7-point Likert scale (see Figure 10.1) 61.2 percent of the managers found informal information transfer with other firms important in some degree (I took '4' as 'not important, not unimportant' and 5–7 as important in some degree). No distinction is made between minimills and specialty steel firms. According to von Hippel (1987) Quanex, the outsider, qualified know-how trading in the specialty steel industry as unimportant or even non-existing. In von Hippel's study minimills were said to select trading partners very carefully; only when one expects the other firm to be an interesting source of information for one's

Table 10.3 Hypothesized influence of discussed variables on employees' decision whether to transfer information to colleagues from other firms

Variables	Likelihood of a transfer of requested information if variable takes on high value
Influencing the costs of transferring information:	
• Degree of competition between information providing firm and information receiving firm	Less likely
• Availability of alternative information sources to information receiver	More likely
• Information's impact on domains of high competitive importance	Less likely
Influencing the benefits of transferring information:	
• Expected change of information receiver's willingness to provide information	More likely
• Value of transferred information to information receiver	More likely
• Technical expertise of information receiver	More likely

Source: Schrader (1991)

own process development problems is information given away.

Table 10.3 gives additional information from Schrader's study. The managers were asked what the likelihood was of employees deciding in favor of informal information transfer when certain conditions were present. These conditions concern the degree of rivalry between the information trading partners, the availability of alternative sources, the strategic value of the information for the receiver and the provider, the probability that the receiver provides information, and the level of technical expertise of the receiver. One was asked what the likelihood of providing information was when the mentioned conditions took on a high score.

As one would expect, information transfer is less likely to occur when the providing firm sees the firm asking for information as a strong competitor and when the information is assumed to have a strong impact on the domains that are important for a competitive position in the industry. Likewise, the more alternative sources the providing firm assumes present, and the more it expects the receiving firm to provide information in reaction, the more willing it is to provide the information. And as was shown in von Hippel's study indeed the firm is more willing to provide information the higher the estimated level of technical expertise of the receiving firm. Perhaps more surprising is that the more important the information is supposed to be for the receiving party, the more likely it becomes that the the firm provides the information. It is difficult to imagine information important for the receiver that is not within the 'domains of where competition is fought'. Apparently there is some trade-off between giving away important information and the likelihood that one receives important information in return.

Schrader also asked to what degree one participates in the informal infor-mation transfer process in the industry and the degree of perceived economic success due to these transfers. He does not give statistical results, but the two seem positively correlated.

The conclusion from the minimill case is that informal information exchange as a loose form of horizontal cooperation is quite significant in this specific industry. One is willing to provide information about process devel-opment if this information is not the core of one's own competitive position and if the rival firm is able (technical expertise) and willing to give equally important information when asked to.

The MCC case

The 'Microelectronics and Computer Technology Corporation' was founded in 1982 as one of the largest private research joint ventures in the world. There were 10 founding members and in 1986 there were 21 large firms from the computer industry, the semiconductor and electronics industry, the aerospace industry, and some other hightech industries were members of the joint venture. Peck (1986, 219) describes the goal of this technological cooperation as 'advanced long term research and development in the fields of microelectro-nics and computer development' which is to provide the members with a new technological base from which they, individually, can undertake product deve-lopment projects. Ten firms built a separate research facility – the MCC joint venture – in Austin in 1983. Each firm paid an equal share of $150,000. For 'latecomers' of the 21 in 1986 the share grew periodically up to $1,000,000. By 1986 four research programs existed and each member firm had to pay a fee to participate in such a program (see Table 10.4).

The yearly budget of the 21 firms in 1983–84 varied from 95.2 to 746.0 million dollars, three of the firms joined all four programs and ten of them only one.[2] To join a program a participation fee has to be paid that is conside-rably higher than the initial share of $150.000 in the joint venture. As may be read from Table 10.3 each program aims at generating technological knowl-edge that forms the basis for whole new generations of products and processes. According to Peck product development in both the computer and chips industry originates from basic technologies that were realized many years and sometimes even decades before. These industries now face the challenge to

Table 10.4 The MCC research program

Program	Number of companies participating	Programme duration in years
Semiconductor packaging To improve methods of placing circuits on microchips and to improve connections between microchips	10	6
VSLI/Computer-aided design To improve the use of computers to design microchips and control their manufacture	12	8
Software technology To develop methods to reduce the costs of writing software	10	7
Advanced computer architecture To improve the design of computer systems (this program consisted of four subprograms)	8	10

Source: Peck (1986, 220)

renew this technological base. Firms in an industry require basic research (or better generic research according to our definitions in Chapter 1) to establish such a new base. With the continued pressure from the market to come up with new products (and new processes to support them) firms in industries like computers and semiconductors can as a rule only spend arround 5 percent of their yearly budget on basic research (Peck, ibid., 221). Compared to IBM and AT&T which have yearly R&D budgets of approximately \$2.5 billion, even the largest firms in the MCC cannot set up competitive basic research programs on their own. In addition the Japanese government stimulated and co-financed similar programs in the early 1980s in which most Japanese competitors cooperated. Similarly the EU came up with ESPRIT and other research programs in which all big European-based firms like Siemens, Philips, and Olivetti participated (see Chapter 11).

According to Peck (ibid., 226) the planners of MCC faced three significant problems in devising its organization: (1) governance; (2) the distribution of costs and benefits; (3) a mechanism to transfer technology from MCC to the members' R&D units. We will discuss these three points further, as the detailed design of the whole organization is in sharp contrast with the 'loosely organized' information exchange between the minimills.

The governance of the joint venture can be divided into two levels. First the corporate level, where a board of directors decides about the research strategy concerning the four programs agreed upon from the outset. For example the board monitors whether the financial implications of the projects undertaken in each program are within the 'ceilings' as set within the package of technological agreements at the foundation of the MCC. Each member of the MCC has a representative in this board of directors and this board appointed an outsider as its chairman and as MCC's CEO (a high-ranked officer retired from the US Navy). At the second level, that of each individual program, program participants decide about the program director and other key personnel. They are also responsible for the budget and technical plan of the program, for incentive payments to program personnel, the payments for latecomers (the acceptance as such is in the hands of the board), and a licensing plan and royalty rate for the use of MCC research results.

> These provisions are intended to make the research program responsive to the needs of the participants who pay for the research. This has been an important consideration in the design of MCC, for many of the members were concerned that existing academic, government, and industry-wide R&D organizations have pursued research that promotes the reputation of researchers among the scientific community at the expense of the needs of industry. [this description is consistent with the definition of basic research, in contrast with generic research, given in Chapter 1] ... MCC management has been designed to be simple and non-bureaucratic. There are only three levels – (1) researcher, (2) vice president and program manager, and (3) chief executive officer. Considerable authority is delegated to program managers, and the ease of access of researchers to managers is emphasized. (ibid., 227)

The division of costs and benefits is a crucial aspect in any cooperative form that goes beyond the loose and informal arrangements like the minimills case. 'MCC has adopted a simple rule for assigning R&D costs: initially, all the participants in a program pay an equal share. (Participants who join later may pay a higher price ...) Obviously, participants will not benefit equally, but it is difficult to predict the distribution of benefits among participants' (ibid., 227). The general idea is that it is up to the individual firms how to profit from the research results of the joint venture in terms of new product and process developments. Rules need to be set as to how research results from the MCC become available. The participants in a program recommend to the MCC board a licensing plan each time 'transferrable' technology has been created. Included in the plan is the precise form of licensing that gives the participants the right to license technology in the first three years in which it is available. This provides program participants with a head start of three years over other members of the MCC, who then have the right to obtain a license at royalty rates set by the participants. Only with the approval of the board is it possible for a program manager to license a technology package to a member outside the program earlier than three years' or even to non-MCC members. 'There is then the possibility of substantial license income which would reduce the cost of program participation to a modest amount'. (ibid., 228)[3] This license income is distributed according to very precise rules. Five percent goes to MCC, five percent to the program researchers, 27 percent to the participants in equal shares as technology credits,[4] and 63 percent as a cash payment to the program participants.

'Technology transfer from MCC to the participants is critical, not only to the overall success of MCC, but to the benefits obtained by individual participants'. (ibid., 228) To facilitate technology transfers quarterly briefings are organized for the program particpants and each of them is allowed to appoint one liaison representative. A liaison representative is an employee of one of the participating firms who spends three-quarters of his time at the Austin laboratories. This gives him the opportunity to brief the researchers at the firm's own laboratories about research results before a technology package becomes available. Each particpating firm to an MCC research program thus has the opportunity to anticipate final results by already planning development projects.

The staffing of the MCC laboratories is remarkable. At the beginning of 1986 65 percent of the researchers were directly hired by MCC. For the remainder 18 percent consisted of liaison representatives who normally spend three years at MCC, and 17 percent of employees from MCC members who remain on the firm's payroll (and MCC is reimbursing the firm for the time the researcher works at MCC). 'Of the seven program directors, four came from non-member companies, two from universities, and only one from a member company'. (ibid., 229) A large majority of researchers and managers of the MCC come from the outside. This is understandable in the light of the desire to establish a new technological base. Apparently the inability of each individual firm to set up a generic research program is not only a question of lack of money, but also a shortage of in-house researchers with the proper technological expertise. One of the big advantages of MCC is its attractiveness for 'basic' researchers to work in specialized fields for a relative long period of time.

3 A THEORETICAL EXPLANATION

Earlier in the book two main ways of looking at technological competition have been distinguished: homegeneous and heterogeneous markets. We argue here that they also make a difference in explaining horizontal technological cooperation. Let us begin with the homogeneous market.

Let us assume there are N firms in a homogeneous market that produce product p1 with technology T1. Each firm has a market share of 1/N and a profit margin PM. Let us further assume that they all undertake an innovation project aiming at the development of a new generation product which can be produced with the old technology; we call this innovation project of each firm $P_n(p2,T1)$. Each firm has an equal but unknown probability to successfully complete the project at time t_{in}, which depends on the expenditure on R&D. In the spirit of what was meant with a homogeneous market the danger exists of 'the winner takes it all'. This is translated here into the assumption that the first firm to innovate doubles its sales every year, that it is 'the only one with a new product' and that all the other firms imitate the new product with the same speed. Imitation time is half the period of innovation, in the sense that the accumulation of the required technological knowledge goes twice as fast, once a competitor has marketed a new product. Hence, all the imitators follow after imitation time $t_{im}=1.5t_{in}$.[5]

Why would a firm want to cooperate in such a situation? According to the time-cost trade-off curve, each firm has the option to speed up its innovation process compared to its competitors, but this would increase total project costs (and technological uncertainty probably rises too). In a homogeneous market the speed of innovation is essential to the appropriability of innovation profits. At the same time market uncertainty can be interpreted here as the chance that competitors are also speeding up their innovation process. Hence, accelerating innovation bears the danger of higher costs, but constant market share and profit margin. To reduce the increase of both technological and market uncertainty due to speeding up innovation, one can cooperate with one or more competitors. The more competitors join, the lower the costs, given a chosen speed of innovation, and also the lower the chance that a competitor (group) is faster. This case is quite similar to the minimills case.

Baumol (1991) has analyzed the minimills case. One of the main questions is whether a firm tries to get more, or more valuable, information from the other firms than it 'transmits' to the others and how this could be detected. As we have seen engineers exchange information without official approval of management, which already makes it difficult for one firm to establish a strategy of cheating. In addition all firms are using the same basic process technology, so engineers of each firm are informed about the kind of problems of the others. This makes the probability high that cheating will be discovered.

As the process engineers of all firms form a contact network, they will all know soon when one firm cheats. The consequence would be that this one firm is excluded from the information exchange. What would be the impact of such an exclusion? Baumol (ibid.) shows this with the following model.

Assume that there are originally N firms in the exchange network and that one firm is excluded from any information. All firms undertake R&D which is directed towards productivity increase. Let us assume that each firm spends X on R&D each year, and that this leads to a unit cost reduction of 1 percent per year. In addition to results from their own research, they can gain an extra 0.3 percent unit cost reduction from receiving information from another firm. Exchanging information with all the firms in the network therefore results in the following cost reduction per year:

$$C_1 = [1 - 0.01 - (N - 2)\, 0.003]C_0 + X/Q_1 \qquad (10.1)$$

Q_1 is the production volume of each firm per year. N-2 is the total number of firms with which information can be exchanged. For the single firm outside the network the unit cost reduction would be:

$$C_1 = (1 - 0.01)C_0 + X/Q_1 \qquad (10.2)$$

When $X/Q_1 = 0.005$ and N = 12 the unit cost reduction for a network firm after one year would be 3.5 percent, while the isolated firm would only realize a reduction of 0.5 percent.

Two questions need to be addressed in relation to what we have dicussed in the book until now. One, what can firm A learn from firm B when they are so similar? Two, what about learning that is communication costs?

As we argued before, two firms that have quite similar technology will communicate relatively easily, but one can question whether they have much to learn from each other. In the case of the minimills, who all have the same process technology, apparently a series of small improvements in process technology are possible. In practice a number of improvements may be aimed for simultaneously, others will depend on previously realized improvements. If we assume that on average L improvements are tackled in each firm's R&D program per year and M improvements in the whole network (M>>L), firm A can maximally learn from firm B when they undertake the same set of improvement projects {Pi} during the whole year, but in an 'opposite' sequence.

It is quite plausible that in terms of Cohen and Levinthal (1990) the improvement projects are all 'close to what the firms already know'. In this sense communication about a project that one firm has already undertaken and the other is in the middle of will not be very problematic; learning costs are expected to be low. However, the more firms in the network (N is very large) the more communication costs (including misunderstandings) may arise. Learning costs and communication costs are both dependent on the number of firms and the number of improvement projects; they could be included in equation (10.2) in the same way as R&D expenditure X.

In a heterogeneous market things may be quite different. When firms in a market are specialized in technologies and each of the technologies in isolation or in combination with one or more of the others can be used to arrive at new products and processes, much is learnt due to cooperation.

Let us take the extreme case where a market in a certain period has N firms and N technologies. Each firm has one technology and it can make one unique product that has to compete with the products from the other firms. From the customer's perspective the products are similar in that they fulfill the same need. Each firm i in the market can therefore be characterized as $\{p, T_i\}$. As we assumed that each product is technically completely linked to a single technology a firm can only gain market share if its technology results in a preference by the customers of its product $p(T_i)$ above the rival's product $p(T_{i'})$, on the basis of lower production costs or a better performance. Cooperation is only attractive for the 'weaker' firm, but it has nothing to offer in return. If however we take the same basic assumptions for this market, but add a segmentation (for instance geographically) things already become different. Assume that each firm has a complete segment of the market S_i that cannot be entered by another firm because of too high transportation costs. Cooperation is an option if the firm characterized by $p(T_i)$ can gain access to $p(T_{i'})$'s market by offering technology T_i. Some of the cooperative technological agreements discussed in Chapter 8 had this market access aspect.[6]

A more attractive initial condition for technological cooperation arises when there is no segmentation and a new product p' can be made with a combination of T_i and $T_{i''}$ (where one technology is not preferred above the other by the customers). When p' is not competing with p both firms could set up a joint venture to develop and produce the new product. To the extent that combinations of other technologies do not result in competing products the development of p' is not so much a matter of speed but of learning costs. How expensive is the development of p' compared to, for example, $p(T_i)$ and $p(T_{i''})$ given the 'different worlds' in which the researchers specialized in T_i and $T_{i''}$ respectively are operating? One could think of two members of the MCC from the computer industry, each with its own technology, now working together on a new generation of technology that is immediately commercialized into a joint product. The problem is not to transfer technology to the separate firms, like in the MCC, but to overcome problems of mixing the two technologies into a fruitful combination. Henderson and Clarke (1990) made clear how the development process of each company might be tuned to the 'architecture' of the existing product. The joint venture can be expected to need an R&D department different from each of the departments of the partners.

The explanation of Teece

In Chapter 4 we discussed how appropriability conditions, the life cycle of technology in the industry, and control on specific complementary assets determine to what extent the innovator in the industry receives all or a large part of the profits from the innovation (see Teece, 1986). When in the industry a weak patent protection exists, and when a dominant product design has been established (paradigmatic stage of the technology life cycle), the best 'guarantee' for the innovator is full control over all specific complementary assets. Normally this implies the need on the innovator's side for production and distribution facilities that can meet the competition of the best firms in the industry and full control over all specific supply units. Full control means full ownership or at least a majority of the shares of the supply firms. According to Teece (1986) such full control is financially impossible in most cases, but we add here that it may also be impossible to always select and own the 'right' suppliers. Because of these financial and rational constraints, firms are 'forced' to cooperate with suppliers, and sometimes even with competitive distributors and manufacturers. In that respect technological cooperation is seen by Teece as a second best solution. The best firms in the industry have competitive manufacturing and distribution facilities, and will own the strategically most important suppliers. It must be stressed that this negative view on technological cooperation is relevant for industries where a dominant design exists and large-scale production, with machines specific for the industry, is the rule. According to Teece (1986) such industries have rather mature technologies; an example is the automobile industry.

In industries with rapid technological change things may be different. The MCC made clear that when a new technological base in the industry is required cooperation in (basic) research may offer considerable advantages, especially as the appropriability of innovation profits for the individual firm is not directly influenced. On the one hand basic, or better, generic research usually has non-anticipated commercial applications, and on the other hand cooperating firms have the ability to commercialize the results of joint generic research individually.

Firms are specializing more and more in technologies, and the number of technologies with which products for various markets can be made increases too. Moreover technology development and product development becomes more expensive every year. Consequently firms must enter geographically large markets to earn back these development costs. This is the story of 'market room' discussed in Chapter 5, but now complicated by the introduction of multiple products and multiple technologies. According to Teece (1992, 2) 'the intensely competitive environment in which high-tech firms find themselves, coupled with the global dispersion of productive technical competence, often requires complex forms of cooperation among competing firms'. Teece states that four elements are basic to the informational and coordination requirements of innovation that ultimately determine whether specific forms of cooperation, full integration or market relationships are the most efficient. In many cases the advantages of technological cooperation are obvious. The four elements are:

1. Accessing complementary assets. These are the production and distribution facilities and component technologies required to commercialize an innovation, discussed in Chapter 5. Here cooperation may undermine the innovating firm's ability to appropriate 'all' innovation profits.
2. Coupling developer to users and suppliers. In the previous chapter it became clear under which circumstances 'early supplier involvement' is needed to speed up or improve the firm's product development process. Equally important are customers, as von Hippel (1988) already stressed. The point here is that the innovator who cooperates most effectively with important customers and suppliers is in many cases the most successful.
3. Coupling to competitors. 'Horizontal linkages can assist in the definition of technical standards for systemic innovations'. (Teece, 1992, 7) In addition agreements with competitors can overcome appropriability problems that will arise when control of complementary assets is too costly. 'The effect of greater appropriability is to encourage greater investment in new technology'. (ibid., 7) Horizontal coordination and communication that is strongly facilitated through cooperation also avoids duplication in R&D. Otherwise diversity in technological development may be too high. Connecting to point two: cooperation with a rival may also provide access to technologically advanced suppliers and customers.
4. Connections among technologies. 'Particular technological advances seldom stand alone. They usually are connected both to (a) prior development in the same technology, and (b) to complementary or facilitating advances in related technologies. In addition, a generic technology may be capable of (c) a wide variety of end-product applications'. (ibid., 8) When an innovating firm has no expertise in prior development of the technology, in related technologies or in (specific applications of) generic technologies, horizontal cooperation and or 'diagonal' cooperation becomes neccesary.

In this way the whole range of possible motives for technological cooperation is covered. It is important to realize that a number of conditions must be fulfilled to become a successful innovator. One condition is access to complementary assets, once the innovation is realized. In this respect the main disadvantage of horizontal cooperation is a weaker appropriability for the innovator. The emphasis in Teece (1992) is on the conditions to come up with an innovation. Horizontal cooperation has some direct advantages here, related to standards for systemic innovations, more control of 'all' complementary assets, avoidance of duplication, and access to previous, related or generic technologies which make the probability of technical success of the innovation higher. The indirect advantages of horizontal cooperation rest on access to the technological competences of the partner's suppliers and customers.

4   CONCLUSIONS

A major concern for firms considering horizontal cooperation is a loss of monopoly profits. Once they enter such a cooperative agreement the danger is that the rival firm gets more out of it. The MCC case made clear that a reduction of (basic) R&D costs can be an important motive for horizontal cooperation. In the minimills case it was the reduction of unit production costs of steel-making as a result of incremental process innovations that mattered. In both cases the threat of large (blocks of) competitors was a strong driving force to initiate the cooperation. From the more theoretical perspective the minimill case can be seen as an example from a homogeneous market, while the MCC case bears elements of a heterogeneous market. The information exchange between the minimills can be interpreted as an effort to collectively improve 'production assets' compared to the large steel firms. In the case of MCC the danger of loss of control on complementary assets is small, because the cooperation concerns 'pre-competitive research' and thus does not change the 'balance of power' between the members (and others in the industry) with regard to complementary assets. The MCC case can be seen as an opportunity to gain economies of scale (per program) and to a lesser extent economies of scope (between the four programs) in R&D. Economies of scope relates to the matter of technological specialization of firms within an industry, that makes horizontal cooperation attractive.

NOTES

1. The largest one is Eastman Kodak (746.0), followed by Digital Equipment Corp. (630.7), and Boeing (429.0). See table 2 in Peck (1986).
2. In fact Quanex, the exception in the study of von Hippel, qualified itself as a specialty steel producer that uses the 'mimimills' process technique.
3. I have not found a text explaining the rights of non-MCC members further. I interpret Peck's explanation as the option for program participation to sell, upon approval of the board, the technology to a non-MCC member only after three years.

4. These technology credits can be used as partial payment on other MCC programs. However, if the firm leaves the joint venture the credits revert to the corporation.
5. For simplicity it is assumed that all firms but the 'winner' have been confronted with a failed product development project once the new product is marketed by a competitor.
6. In this simple case cooperation seems only attractive for one of them, depending on who is the most successful in $p(T_i')$'s market with the same technology T_i. One could imagine that the size of this market segment increases due to the 'better' technology and both firms work out some sharing arrangement. Note that no economies of scale can be gained through cooperation, due to the high transportation costs.

REFERENCES

Baumol, W. (1991), Technology-sharing cartels, Paper presented at the EARIE conference, Cambridge

Cohen, W.M. and D.A. Levinthal (1990), Absorptive capacity: a new perspective on learning and innovation, *Administrative Science Quarterly* 35, 128–52

Henderson, R. and K. Clarke (1990), Architectural innovation: the reconfiguration of existing product technologies and the failure of established firms, *Administrative Science Quarterly*, 35, 9–30

Hippel, E. von (1987), Cooperation between competing frims: informal know-how trading, *Research Policy* 16, 291–302

Hippel, E. von (1988), The Sources of Innovation, Oxford University Press, Oxford

Mariti, P. and R.H. Smiley (1983), Cooperative agreements and the organization of industry, *Journal of Industrial Economics* 31, 437–51

Peck, M.J. (1986), Joint R&D: the case of Microelectronics and Computer Technology Corporation, *Research Policy* 15, 219–31

Schrader, S. (1991), Informal technology transfer between firms: cooperation through information trading, *Research Policy* 20, 153–70

Teece, D.J. (1986), Profiting from technological innovation, *Research Policy* 15, 286–305

Teece, D.J. (1992), Competition, cooperation, and innovation: organizational arrangements for regimes of rapid technological progress, *Journal of Economic Behavior and Organization* 18, 1–25

11. National systems of innovation

1 INTRODUCTION

In the last three chapters technological cooperation has been discussed. Today the role of government in promoting technological cooperation is part of what is now called the 'national system of innovation'. However, the national system of innovation includes much more than cooperation between private firms and government's efforts to promote this cooperation.

Earlier in the book it has been made clear that innovating firms often use external sources of (technological) information. These sources can be other private firms as well as public organizations. The latter can be said to belong to the 'public technological infrastructure' of a country. Innovating firms may derive technological knowledge from external sources of information. More generally firms interact with several kinds of outside partners that co-determine the initiation of innovation projects or the execution of existing ones. Consequently each innovating firm can be seen as participating in one or a number of innovation networks. Such innovation networks may both cover formal co-operative agreements and informal information exchange relationships. At the same time public and private organizations are involved. Besides innovating firms, public research institutes, public organizations concerned with technology transfer towards private firms, and private consultancy firms offering specialized services with regard to innovation and technology transfer may participate in such innovation networks.

We have seen that technological cooperation can be divided analytically into horizontal, vertical, and diagonal relations. The difference between technological co-operation as discussed in Chapters 8–10 and innovation network is twofold. It concerns the number and kind of participants. Technological co-operation can be between two firms, while a network is between three or more partners. Technological cooperation – as discussed until now – was between private firms, while in innovation networks private firms, public research institutes, and other organizations, for example innovation centers, may be involved. Seen from this new angle one may expect 'horizontal', 'vertical', and 'diagonal' networks, with or without public organizations, and networks with a combination of the forementioned possibilities.

There may, however, be additional organizations and other institutions which facilitate innovation, that were excluded from technological cooperation but could be added to innovation networks.[1] Some have already been discussed or at least mentioned, such as the Patent Office, national and regional government organizations for technology policy, chambers of commerce, innovation centers, private and public venture capital organizations, other financial institutions involved in the finance of innovation, and so on. A system of innovation can be defined as a group of private firms, public research institutes, and several of the 'facilitators' of innovation, who in interaction promote the creation of one or a number of technological innovations. The definition of innovation system can be extended by including institutions which promote or facilitate the diffusion or application of these technological innovations.

We have seen that basic research is largely undertaken by, or at least financed by, government. There is a relation between this public basic research and R&D of private firms: knowledge created in basic research will on many occasions be used as an input to innovation projects of private firms. This will quite often occur many years later or in another direction of application than expected, in another country than the original basic invention, and so on. More generally a considerable amount of public scientific and technological knowledge is available in society. We will call the set of institutions which possesses such knowledge or which occupied with the transfer of that knowledge to private firms the public technological infrastructure. The interest of economists and other researchers into the question of how this public knowledge can be transferred more effectively and efficiently towards private innovating firms has grown over the last decades. This is presumably due to the considerable increase in the speed of innovation with which private firms need to fight international competition on their markets.

The public technological infrastructure is a part of the national innovation system in a country. In this chapter we first describe innovation networks and innovation systems on the base of empirical research into the matter. Then we proceed, as usual, with some theoretical aspects. The core is interactions among the various actors in the network or in the whole system, and one of the main aspects is knowledge or technology transfer.

As already mentioned in the previous chapter current attention is drawn towards the ways in which governments try to promote 'technology transfer' from the technological infrastructure to its private firms. In the empirical section some new practices of technology policy in Europe are discussed. In the theoretical discussion we focus on 'the fundamentals of new technology policy', including the current tendency in many European countries 'to let the market work'.

The problem

In the previous chapters technological cooperation between private firms has been reviewed. An innovating firm has a whole array of possible sources of innovation. By now it is common in the innovation literature to distinguish between private and public sources of information for innovation. Public sources include governmental research institutes and university research laboratories. Every country or region may possess a set of such public 'knowledge centers'. The main problem is that the kind of knowledge created in public institutes is not suited, or at least not ready-to-use, for private innovating firms. A knowledge transfer problem often exists that prevents firms from exploiting all kinds of public information regarding technological matters. Therefore public or semi-public 'transfer organizations' have been established in many industrialized economies. Public research institutes and transfer organizations together are important elements of the national system of innovation. The better the technological infrastructure is organized and 'custom-made' to the firms' innovation requirements the better the innovation performance of the private sector will be. While we already saw how cooperation may influence R&D productivity with given levels of firm R&D expenditure, the technological infrastructure adds to this new insight: the more a firm, an industry, a region or a country exploits its public technological knowledge the higher will be R&D productivity at those levels of analysis. This chapter focuses on the relations between private firms and 'knowledge centres' and more broadly with the interactions between these firms within the national system of innovation. The examination includes government's role in promoting these interactions. What is also included in most descriptions of national system of innovation are the relationships between innovators and their customers and suppliers, as far as they are within the boundaries of the innovation system. In this sense a national system of innovation is an important cornerstone to the competitive advantage of nations (Porter, 1990).

While technological cooperation covered relations between private innovating firms with their competitors, suppliers, and customers, the current chapter shows the modern role of government in promoting cooperation and knowledge transfer. The question is what do 'systems of innovation' add? The answer is that national innovation systems include more organizations and institutions and focus on interactions within the national territory. One of the main kinds of institution included is education relevant to innovation and the diffusion and application of new technology. Other institutions we have already mentioned.

As already suggested in the foregoing, there may be regional or sectoral innovation systems as well. Regional or sectoral innovation systems are subsystems of the national system in which the institutions (or some of them) are specialized in the innovation problems of a specific sector or region.

According to Cohen and Levinthal (1990) R&D is the firm's main mechanism to learn from outside sources. In Chapter 3 we made clear that R&D is only one out of a number of various activities needed to carry ideas through to commercial exploitation. Perceiving the national system of innovation as the set of institutions promoting knowledge transfer towards the

Figure 11.1 The national (regional) network in international markets

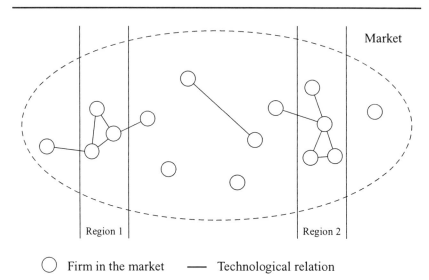

country's innovating firms is far too simple. It should also be seen as including institutions influencing the finance of R&D, the application of patents,
innovation project management, the selection of new technology fields for the
future, and so on.

The emphasis on the national territory follows from the assumption that
innovations created by domestic firms are (much) more the result of the interactions between these firms and other domestic actors within the innovation
system than of interactions with foreign parties. Given the ongoing internationalization of many markets this is an interesting assumption. It implies
that the 'domestic' part of a firm's innovation network(s) is still very important for its position *vis à vis* the other firms in the market. Figure 11.1 reflects
this possibility in terms of number of network relations. More realistically
would perhaps be that not the number but the intensity and effectiveness of
domestic relations is higher.

In the empirical section we examine the structure of the systems of innovation and in particular the way in which innovating firms try to 'exploit' the
technological knowledge from the public infrastructure. The ways in which
government plays an active role in the national system of innovation is discussed too. A particular point is the initiatives of the EU to promote paneuropean cooperation in innovation. As these technology policy programs in
Brussels may ultimately create a European system of innovation they are discussed too.

In the theoretical section we explore some new and tentative insights on public knowledge centers and transfer problems between public and private institutions. The discussion is also about the national boundaries of systems of innovation in an economic world that is increasingly international instead of limited to administrative areas.

From a theoretical perspective the main question remains (see Chapter 6) that if markets do not function properly with regard to the creation, application, and diffusion of new technology, how can governments improve this situation and how does their effort relate to private inititatives (such as technological cooperation) to reduce uncertainty of innovation processes? Does the national innovation system result in an increase of R&D productivity or an increase in the speed and quality of innovations carried out by the domestic firms? How does this relate to the firms' operation in international markets?

2   THE EMPIRICAL SIDE

Innovation networks

In Chapters 9 and 10 examples have been discussed of relationships of an innovating firm with other firms in the same industry and with customers and suppliers. Larger firms usually have a number of advanced development or technology projects and a much larger number of product development projects. Some projects may be undertaken by the firm alone, many others will involve cooperation with 'outsiders' in one way or the other. Seen from this perspective of a set of specific technologies and specific product development projects a firm is expected to be involved in a number of innovation networks simultaneously. Figure 11.2 depicts such a situation.

Network T1 is an example of a 'pure horizontal' network; three firms from the same industry are cooperating, for example in the development of generic computer and microelectronics technology as was the case in the MCC joint venture (Chapter 10). Network P1 is an example of a 'pure vertical' network in which a single assembly firm is cooperating with a number of its suppliers (main suppliers or co-makers). In P2 two firms are jointly developing a new product with the aid of some of their suppliers. In T2 a number of competitors and some supposedly advanced suppliers from one or more of these firms are jointly developing basic technology. P2 and T2 are examples of networks with a mixture of horizontal and vertical relationships.[2] In network T3 three firms from different industries (diagonal relationships) are working together on new technology.

The example suggests that two or more firms participate simultaneously in more than one network. This may strengthen the bonds between the firms

Figure 11.2 Various innovation networks

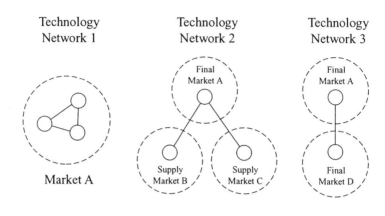

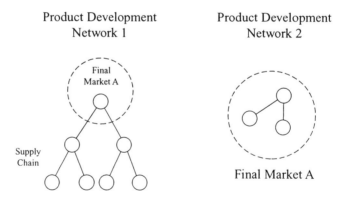

and diminish the need to create a new 'common culture' every time again. The previous chapters provide several examples of horizontal and vertical networks. The cases of the aircraft industry made clear that in practice a mixture of horizontal and vertical relationships within a network exists.

Airbus cooperation is an example where government has been involved in the initiation and finance of the projects. In a number of cooperations public or publicly financed institutions, mostly research institutes, work together with private firms.

Public–private innovation networks

Lundgren (1995) gives an example of the construction and development of the Swedish network 'around' digital image technology during the period 1975–1989. Particularly in such an emerging field of technology a mixture of private firms and public research institutes is quite common. Kreiner and Schultz (1991) provide examples of cooperations in biotechnology where universities and other public research institutes work together with small biotechnology firms. The relationships within such networks of emerging technology are generally more of an informal kind than based on specific contracts. The description of the development of a specific technology, such as 'image recognition', is sometimes understood as a 'technological system'.

While an innovation network is often centered around a specific firm, a technological system is broader; several firms of equal importance are taking part, as well as public research institutes and other organizations influencing the development of the technology. The difference with a national system of innovation is the technological system's preoccupation with a single technology or set of related technologies.

Technological systems

A national system of innovation can be subdivided in various ways: networks around individual firms, industries, regions, and technologies. The subdivision into industries is the most traditional one in economics. A common factor for the firms in a particular industry is the customers for which they compete. A disadvantage of this perspective is that competing firms with a different nature of innovation network are not recognized. In particular the involvement with different technologies is important from the perspective of innovation. A technological system concentrates on firms occupied with the same technology (whether they are competitors or operationg in different markets). Carlsson (1997) discusses four different technological systems in the Swedish national system of innovation, thus providing an important empirical depth to the often superficial pictures at the national level. Figure 11.3 gives an impression of the details of a technological system's approach in terms of actors involved in the Swedish 'Powder Technology' system.

Characteristic of this technological system and others discussed in Carlsson (ibid.) is the amount of various scientific and technological disciplines involved in the (commercial) development of the technology. For the four main problem areas identified for powder technology 20 different scientific and technological disciplines and specialties have been put forward (Granberg, 1997, 176).

Figure 11.3 The technological system of powder technology in Sweden

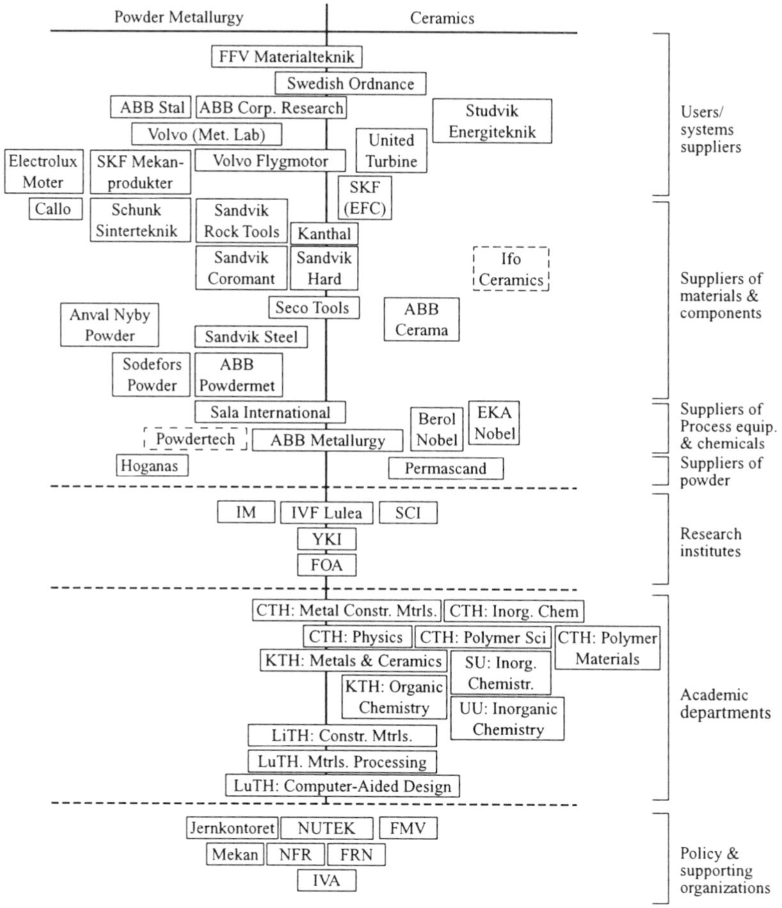

Source: Carlsson (1997, 180)

National systems of innovation

According to Nelson and Rosenberg (1994, 20) the term national innovation system and its basic conception was used and perhaps even introduced in Dosi et al. (1988), where part V of the book was titled 'National Innovation Systems'. Included in that part is a description of the national innovation system of Japan (Freeman, 1988). The following institutions are mentioned in the description of the Japanese system: MITI (and government more in general), the firms and especially the Keiretsu, universities, the informal consultation process between these three players, general education and training, lifetime employment, the relatively small wage differences between management, white-collar, and blue-collar workers, and other social innovations stimulating innovation.

It is very difficult to give a full account of all the institutions in a country and how they work together to accomplish innovations. It is evident that countries differ in the number, nature and interaction of those institutions. Ergas (1987) makes a distinction between 'diffusion oriented' countries and 'mission oriented' countries. In mission-oriented countries, like the US, France and the UK, large government (co-)financed research programs, often related to defense or aerospace, play the role of 'catalyzers' towards innovation in the private sector. Otherwise involvement of government is relatively modest and largely confined to subsidy schemes and tax facilities as discussed in Chapter 6. In the US and UK the 'market for corporate control' is argued to be quite efficient. This means that the stock market is a good signalling device for 'bad management' and mergers and acquisitions are used to correct these failures. There is a tendency for short-term value increase for shareholders in such systems and the capital market is reluctant to finance long-term innovation projects. At the same time, however, the US, and to a lesser degree the UK, are characterized by a well-developed venture capital market. In France government involvement is traditionally much greater than in the UK and US. Apart from the large 'mission' there is a web of regional organizations promoting innovation, especially in small and medium-sized firms. In diffusion-oriented countries like Sweden, Switzerland and Germany, government, banks and (specific) industries are much closer linked. Banks have shares in the most important industries and a strong bond exists between industry and government in keeping up high levels of training and professional education (an example is the mechanical industry in Germany). According to Ergas Japan is a mixture of mission-oriented and diffusion-oriented. His attempt to classify the main industrialized countries from the early 1980s according to some main elements of technological development can be seen as one of the first implicit uses of national innovation systems.

Part of the national system of innovation (NSI) is the organization of R&D. While many descriptions of the NSI of countries lack detail in terms of the various institutions involved, descriptions of the organization of R&D are usually more complete. Figure 11.4 gives an example of the 'research system' in the UK.

This British picture of the organization of research is quite detailed. Most people involved with innovation would be able to draw a comparable picture for their own country. This research system is part of the 'Research and Development' system of a country. Figure 11.5 gives an idea of the various amounts of money in the R&D system in Germany.

Figure 11.4 The UK research system

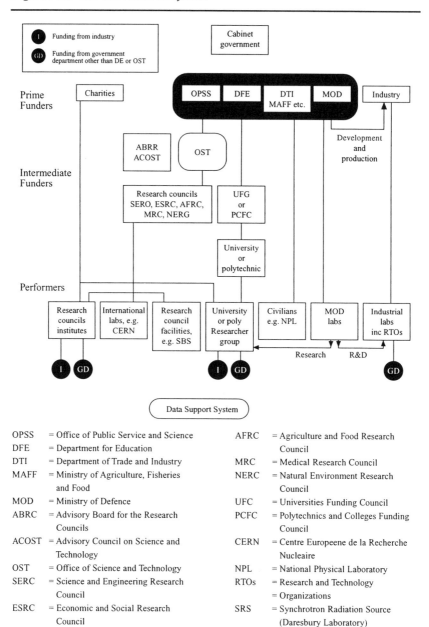

OPSS	= Office of Public Service and Science
DFE	= Department for Education
DTI	= Department of Trade and Industry
MAFF	= Ministry of Agriculture, Fisheries and Food
MOD	= Ministry of Defence
ABRR	= Advisory Board for the Research Councils
ACOST	= Advisory Council on Science and Technology
OST	= Office of Science and Technology
SERC	= Science and Engineering Research Council
ESRC	= Economic and Social Research Council
AFRC	= Agriculture and Food Research Council
MRC	= Medical Research Council
NERC	= Natural Environment Research Council
UFC	= Universities Funding Council
PCFC	= Polytechnics and Colleges Funding Council
CERN	= Centre Europeene de la Recherche Nucleaire
NPL	= National Physical Laboratory
RTOs	= Research and Technology = Organizations
SRS	= Synchrotron Radiation Source (Daresbury Laboratory)

Source: Saviotti and Gummett (1994, 23)

As we argued earlier, the way a country organizes the transfer of knowledge from the public technological infrastructure or the research system as depicted in Figure 11.4 is as important as a highly-qualified research system on its own. The organizations in the UK promoting technology transfers from research to the product development departments of private firms could therefore be added.

The promotion by governments of technology transfer from the public infrastructure towards firms is an increasingly important part of the NSI. The variety of institutions involved, like the Chamber of Commerce, regional authorities, the economic departments of large cities, innovation centers, programs for in-company projects for students, branch technology centers, transfer points at universities, makes it almost impossible to provide descriptions of a country's NSI at this point. Moreover to understand the working of such institutions requires deep insight into an even broader range of institutional structures and processes, that are mostly quite country-specific. A further description will therefore be limited to the Dutch situation that is most familiar to the author. Two institutional elements will be discussed: the operation of innovation centers and the recent change at TNO, the largest technology

Figure 11.5 Financial flows in the German R&D system

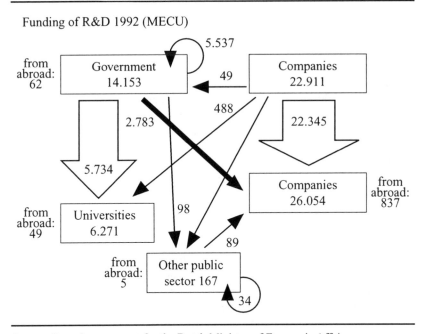

Funding of R&D 1992 (MECU)

Source: Consultancy report for the Dutch Ministry of Economic Affairs

institute. Both are exemplary for the recent changes in many countries that are to a certain extent contradictory: first the recognition that knowledge transfer from the public infrastructure towards firms, especially small and medium-sized ones, needs support; second that this support needs to use 'the market mechanism' as much as possible.

In 1988 and 1989 a total of 18 innovation centers were founded in the Netherlands. The aim of these centers is the promotion of the utilization of technological knowledge by small and medium-sized firms (SMFs) in their region. Initially the centers focused on knowledge available at public research institutes and technical schools, but today also large firms are addressed. The establishment of the centers was financed by the Ministry of Economics that also provided part of the personnel, coming from a more centrally organized institution involved with (smaller firms in) the manufacturing industry. It was planned that the financial support by the ministry would be gradually diminished from 100 to approximately 50 percent. This forces the centers to provide more of their services for the SMFs at a (commercial) fee or with support of other organizations. The services are, among others: direct advice concerning the management of innovation and all other organizational matters related to innovation;[3] establishing links with specific technology institutes and the like that could solve problems articulated by SMFs; supporting SMFs in articulating their technological problems; and organizing meetings for larger groups of SMFs concerning specific topics such as the use of project management methods, the possibilities of a new material, the application for R&D subsidies. At the base of these services is the establishment by the innivation centres of an extensive set of networks both with SMFs in the region and the public infrastructure. Such networking tasks, that must be updated continuously, can only be undertaken with government subsidy, unless fees for SMFs rise to a 'full' commercial level, including overhead costs. It is well known that SMFs are very reluctant and have limited resources to cover full fees for external advice.

TNO is a chain of technology institutes covering most fields of 'natural science'. Their aim is to undertake applied research and to offer their knowledge on specific domains to SMFs in support of innovation. Originally the TNO institutes were completely financed by government, but in the 1990s subsidy levels have been reduced. Consequently the institutes must sell part of their knowledge to SMFs in order to survive. In this way the institutes are stimulated to organize and exploit their applied research in such a way that SMFs become more interested in paying for advice and buying specific knowledge.[4] Of course intermediary organizations like the innovation centres could still play the role of linking pin.

By reducing public support for both the innovation centres and the TNO institutes government is less 'sensitive' to the question as to whether efforts to promote the diffusion of technological knowledge are efficient. The efficiency is partly revealed by the number of contracts the institutes manage to get. Nevertheless the question of efficiency remains relevant, also because a large number of local and regional public organizations is involved in supporting SMFs and both the costs and effects of these simultaneous efforts are difficult to estimate.[5]

Another element of the NSI is the promotion of technological cooperation by government. We discuss this element as it is a 'logical' extension of the private forms of cooperation discussed in the previous chapters.

The promotion of cooperation by government

In the late 1980s policy makers slowly began to recognize that technological cooperation between firms might be important for the competitiveness of their economies. While the large number of cooperations that we know now began to emerge it was also realized that the pattern and extent of cooperation as they were 'coming out of the market' need not be acceptable for specific economies. For example the tendency of large manufacturing firms towards outsourcing in a specific EU country could result in a loss of production, when domestic suppliers are not able to meet the requirements of these manaufacturing firms. This danger of losing part of the production was sharpened because non-fiscal barriers between the EU members slowly diminished and more and more foreign firms tried to penetrate formerly 'closed' domestic markets. It was also recognized that the innovativeness of the economy might be stimulated by domestic as well as international cooperation.

As a result an increasing number of industrialized countries' governments began to promote cooperation. In an analysis of technology policy in eight European countries in the early 1990s Limpens et al. (1992, 28) conclude with respect to international collaboration

> most governments have a separate budget for it, and they attach great significance to it. Although one could expect that the smaller countries would be more interested in this than the larger ones, this does not hold true. The Netherlands, Norway, Switzerland and Belgium are certainly among the countries which contribute strongly to international programs (like the European space activities), Germany is also investing heavily in this area.

Earlier countries began to include instruments to promote domestic cooperation in their technology policy programs. For example, in the US 'the government has cleared the way for collaborative research through the National Cooperative Research Act of 1984' (Niosi, 1994, 3), that made technological cooperation no longer 'automatically' in violation with Antitrust Law.

Technological cooperation among European firms has also been the main motive for the European Committee to start the Paneuropean technology programs, such as ESPRIT, in the early 1980s. We present some figures about these programs and explain how they work.

In fact a distinction must be made between the EUREKA project and the other projects such as ESPRIT, BRITE, RACE, and JESSI. The difference is twofold. EUREKA is directed towards commercialization of new technology, while the other projects officially are aiming for 'pre-competitive' research, comparable with the generic research undertaken by the MCC (Chapter 10). Although the EU has been involved in the EUREKA initiative, the organization and execution of the program is an intergovernmental mechanism. Firms undertaking projects are subsidized by the national governments and the rules of application and funding of a project in which several firms from different member countries are involved may be different, depending on the firm's national government. Firms in ESPRIT and the other projects are subsidized by the EU. They are part of the so-called 'Framework Programme of Community Activities in the Field of Research and Technological Development', that is the major instrument for EU's technology policy. The first Framework Programme was from 1984–87. Meanwhile the fourth Framework Programme (1994–98) has almost ended. The Framework Programme consists of a number of 'specific programs' which describe the technological topics and procedures for carrying out the programs. Although the number and nature of the specific programs changed regularly, the main topics covered are the same as in most countries' technology policy programs: Information Technology, Biotechnology, New Materials and Energy. A separate topic is 'Human capital and mobility' that aims at stimulating the diffusion of knowledge and coordination of research efforts within Europe. It could be seen as part of the underlying motive of the Programmes: promoting technological cooperation among firms and research institutes within the EU and Europe at large.

The major requirement for a project to be submitted for subsidy within the Framework Programme is that institutes from at least two, but preferably more EU member countries are taking part. Today firms and research institutes from other European countries are also allowed to participate. The quality of the project and the quality of the participants are the main criteria for subsidy. Depending on the number of applicants per specific program the probability of receiving a subsidy can be rather low. According to Reger and Kuhlmann (1995, 20) 'the number of applicants has *risen* [emphasis in original] substantially in the past few years, with the result that the acceptance rates have dropped continuously – in information technology (ESPRIT II) the approval rate in the selection following the last call for proposals was only 20 percent'. The selection procedure is as follows. First, with the help of independent referees the most valuable (in terms of the selection criteria) proposals are selected. This short list is then appraised by the program committee and is passed on to the appropriate Director General of the Commission, who makes the final decision. Proposals are already quite elaborate and on acceptance a standard contract of the EU has to be signed by the major contractor and the other participants in the project. The EU provides a subsidy of maximally 50 percent of the anticipated project costs, but only for the first half year. The major contractor is responsible for the progress reports that need to be written twice a year. Only on the basis of these reports and with sufficient results are subsequent half-year subsidies given.

EUREKA was launched in 1985, partly as a reaction on the Strategic Defense Initiative in the US that opened up large R&D potentials in hightech fields for American firms. In contrast to the EU programs these defense contracts could include the development of products for commercialization. According to Peterson the Eureka program's creation was also motivated by the reluctance of EC member states to give up national R&D prerogatives to the commission when huge increases in EC R&D spending were proposed in 1985 (Peterson, 1993, 243). In contrast to the EU technology programs EUREKA is not focused on specific technologies, such as microelectronics and telecommunication and '[it] is a loose framework with no overall strategic goal and few strict rules of process.' (ibid., 244)

From the private initiatives to cooperation discussed in Chapter 10, it is well known that many are terminated because of a lack of success in the eyes of one or more participants. This, however, may not be seen as too negative, as it was argued that particularly the loose forms were intended 'to find out about potential partners'. Nevertheless also many joint-ventures (an intensive form) remain without success. Generally we know less about the 'innovation output' of cooperations than about the R&D programs of single firms. The same holds for cooperation in EUREKA and the EU technology programs. Most empirical research in the last mentioned area is limited to the measurement of the number (and intensity) of linkages between the private firms involved (Hagedoorn and Schakenraad, 1990; Mytelka and Delapierre, 1987). Peterson (1993) provides results of a survey in which participating firms were asked about the results of government sponsored cooperation and their motives for participating. More than 50 percent of the participating firms in EUREKA expected benefits in the areas of 'cross-fertilization of ideas', 'improvement of competitive position' and 'reduction of R&D costs'. (ibid., 256) No questions were reported that directly asked for (quantitative) measures on innovation output.

3   THE THEORETICAL SIDE

This theoretical section addresses the question why and how government needs to undertake the technology programs described in the empirical section, and if so, whether it is done effectively and efficiently. The discussion is related to the idea presented earlier in the book, that the diffusion of knowledge towards private firms is problematic and certainly not guaranteed by competition between those firms. The construction of transfer mechanisms by, or with the aid of, government may thus facilitate the diffusion of knowledge. A complication is that this diffusion is as much dependent on the abilities of the firms themselves and relationships between them as on the 'quality' of the public infrastructure, including government supported efforts to link the two. In fact the web of relationships among firms and between firms and organizations 'within' the public technology infrastructure themselves can be seen as a neccesary condition for the transfer of (tacit) knowledge. We therefore begin with analyzing the innovating firms again, but now from the perspective of relationships among them.

A firm's innovation network

Innovation networks 'surrounding' a firm and systems of innovation at the level of an industry, a region or a whole country have in common the notion that the realization and economic success of a particular innovation is dependent on the actions of other firms and organizations. This 'interaction' approach to innovation is central in the 'Swedish network school' of which Håkansson (1987, 1989) can be seen as representative.

In this approach innovation networks are assumed to emerge and develop (including disappearing again) as a result of various forces that can be summarized by emphasizing two aspects. First a network at a certain moment is the result of decisions and actions of its participants, regardless of whether these decisions and actions are rational and predictable. Decisions and actions are usually the result of a variety of circumstances and motives that, from a scientific point of view, can be analyzed as different manifestations of the network. Put otherwise these factors determine transactions and transformations that are seen as the core activities that take place within networks, via the relationships within the network. Relations exist between actors, activities, and resources. There are three basic ties between actors: affective, connative, and interdependence (Power) (Håkansson, 1987).[6] The nature of these personal ties among actors may influence the interactions among resources and activities, and the latter types of interaction affect personal ties. Crucial for an understanding of the network approach is that:

'Each one of the three components – actors, activities, and resources – is dependent on the other two. Actors are defined by their performance of activities and their control over resources. Activities are performed by actors, a process during which certain resources are used in order that others should be refined. And, finally, resources are controlled by actors and their value is determined by the activity in which they are used'. (Håkansson, 1989, 16–17)

Figure 11.6 shows the most important links between the three components of Håkansson's network model.

Activity cycles and transaction chains are typical for networks. Through these processes are transformation acts (that create economic value for a specific firm) depending on transactions and transformation activities carried out by others in the network. This implies that the value of an actor's resources is also dependent on the resources of other actors in the network.

The second aspect is a dynamic one. The development of the network follows from a continuous 'struggle' between stability and change. From one

Figure 11.6 The network model of Håkansson

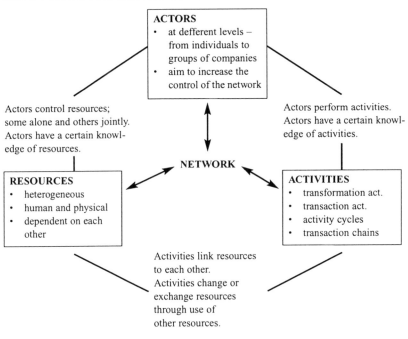

Source: Håkansson (1987)

perspective the interdependencies in the network create stability, as the profit of one actor is dependent on the activities carried out by the others. From another perspective each actor tries to improve his performance. This also means gaining more control over the resources in the whole network, for example by introducing a new product for which the demand outside the network is very promising. This technological change puts pressure on the other actors and may provide the innovator an advantage. We have seen how control of complementary assets is important for the appropriability of innovation profits by the individual firm. Change not only comes from within the network, but also from outside. The more the network is open to the wider environment the more change in this sense is likely to occur. Openness is obtained in two ways. First, each partner in the network may carry out activities and own resources that are outside the network (a firm may produce two products and it only participates in a network with one of them). Through these outside activities firms inside the network may learn additional things. Second, new firms may enter the network and the relationships with others may be terminated.

The transaction cost economic explanation of cooperation and the more strategic analysis by Teece (1986) of this phenomenon were carried out from the perspective of a single firm that has relationships with other firms. A disadvantage of such a 'dyadic' approach compared to a network approach is that the behavior of that firm can only be influenced by firms with which it has a direct relationship. In the example of the aircraft industry in Chapter 9 we saw how CP came in touch with Hispano Suiza through a common business relationship outside the network in which the two ultimately operated. Another important difference is that in TCE and the Teece analysis network relationships follow from (boundedly) rational decision making of firms, given a well-defined set of conditions. These conditions (the three characteristics of the transaction or the three conditions in the environment of the firm) determine the choice of partners and the nature of the relationship with them. In the network perspective described in this chapter the outcomes of the firms' decision-making processes are much more open and so is the structure and development of the network over time.

Why NSI?

Lundvall (1988) and others have taken the network approach one step further. Each innovation network is embedded in a wider web of relationships. Actors, resources and activities in the network are dependent upon and influence in their turn other actors, resources, and activities. On the one hand this means that the earlier mentioned openness of an innovation network inevitably means that each network is connected to other networks of firms. On the other hand an existing innovation network of private firms can be extended by introducing, for example, the connections of the firms with the public infrastructure of the country (or countries, in case of an 'international' network). Descriptions of networks, similar to the one explained above, do indeed include all kinds of institutions other than private innovating firms, as the example of the 'image technology network' showed. And in fact Håkansson's network model does not preclude this, as 'actors' may be innovating firms as well as other organizations that play a role in one or more innovations from the network partners. One can therefore see the NSI as a cluster of interdepending innovation networks.

One of the main virtues of the network approach is that the many complexities concerning innovation (and especially the dependence of one particular innovation on a range of other innovations in the past and in other places) are not harnassed by a specific set of behavioral assumptions about innovators and conditions in their environment. This virtue is at the same time an inherent weakness, for the performance of a particular network is merely the result of the actions undertaken. There is no 'objective' measure such as the R&D productivity of the firm. All innovation network performance is at best comparable with the performance of 'similar' networks. In the case of innovation networks comparison may be difficult, as innovations by definition are unique and each network may have its unique history and institutional context.

Where comparison of individual networks is difficult comparison of different NSIs is awesome, first, because of the complexity of each NSI, and second, because each NSI is in some way related to the internationalization of markets. This may imply that each NSI is increasingly open to other NSIs and all are influenced by the common characteristics of the 'world economy'. The least this implies is that the performance of innovating firms and domestic innovation networks is partly influenced by the national institutional setting and partly by the 'rules of international competition'.

In between networks of innovating firms and the NSI are technological systems. While firms in an industry may fight for the same customers over an extended period of time, a particular technological system may change frequently during a relatively short period. Understanding the dynamics of a technological system, how problems are solved and new problems identified, how actors come and go and how individual actors change their strategies and motives, is the core of the matter (Carlsson, 1997). While the advantage of the 'technological system approach' is the focus on a single technology and its developments, the practical implications (as reflected in the earlier mentioned four empirical studies in Sweden) make clear that a single technology is difficult to isolate from other technologies that are needed to make commercial applications of the technology in often various fields of economic activity (ibid.). As a result the analysis of a technological system, the international connections of its firms and its evolvement over time becomes as complex as the analysis of an NSI. Similar to Håkansson's distinction of the actors in an innovating firm's network, Carlsson (1997) makes the distinction of informal connections between individuals in a nation, supplier buyer relations that are often internationals and the knowledge-creation/problem-solving relations between organizations that are ultimately determining the success or failure of the technological system. The overlap between these 'subsystems' and the difficulty in measuring such relationship characteristics make empirical research problematic, especially when, in the analysis, the system is linked to its technological and economic performance.

Regardless of these methodological considerations, what the network approach and its extension in the form of the NSI have made clear is that the innovation process of the firm is strongly influenced by other firms and organizations. This may have considerable consequences for the analysis of the innovation performance of the firm related to its R&D intensity. The performance depends on the firm's R&D intensity, on the R&D intensities of other firms with which it cooperates, and on the 'effectiveness' of the links with outside organizations (see Palda 1986, for a similar discussion).

From a policy perspective this means that where innovating firms are increasingly burdened by the rules of international competition that are the same for all, the way to gain advantage is by doing better within the domestic networks and NSIs. According to Carlsson (1997) four aspects are crucial in analyzing technological systems: the nature of knowledge creation and knowledge spillover mechanisms, receiver competence (Cohen and Levinthal's absorptive capacity) of the various actors, the connectivity between the actors (where social and geographic proximity play an important role), and variety creation mechanisms (that can be linked to our diversity discussion). A significant addition to our analyses thus far is the assumption that spillover effects in an NSI or technological system are dependent on the nature of technology, although implicitly we have recognized this by the distinction between tacit and codified knowledge and by the problems occurring with architectural innovations.

4 CONCLUSIONS

Innovating firms can be seen as operating in their specific networks. Such innovation networks consist of formal and informal cooperative agreements with other firms. An innovating firm also has connections with the public technological infrastructure of a country, and more organizations and institutions influence the firm's innovation process. A system of innovation that can be analyzed at the national, regional, or sectoral level encompasses a large number of firms and their networks. The European technology programs can be seen as an effort by the national and paneuropean authorities to strengthen the European networks of their firms. From a theoretical perspective the 'network approach' is best able to explain innovation networks, their dynamics, and the relationships with innovation systems. The disadvantage of this approach is that it does not rely on a given set of behavioral assumptions and conditions in the environment, comparable with transaction cost economics or the aproach taken by Teece.

NOTES

1. It is my own idea to make this distinction between technological cooperation, innovation network, and other 'facilitators' of innovation. I do this to distinguish innovation networks from 'systems of innovation' later on.
2. Vertical and horizontal are relative terms. The results of an innovation network may even turn around what was previously a horizontal or vertical relationship.
3. In SMFs this could mean practically everything, because innovation is linked to the whole organization.
4. TNO institutes were often accused of not being able to articulate their knowledge in such a way that SMFs become convinced about its commercial value.
5. One of the points of discussion today is whether a further decentralization of technology policy instruments could improve the coordination and efficiency of all the different actions of the various organizations.
6. But of course the relationships are as much a result of actions and decisions as they are constituting them.

REFERENCES

Carlsson, B. (ed.) (1997), *Technological Systems and Industrial Dynamics, Economics of Science, Technology and Innovation volume* 10, Dordrecht, Kluwer Academic Publishers

Cohen, W.M. and D.A. Levinthal (1990), Absorptive capacity: a new perspective on learning and innovation, *Administrative Science Quarterly* 35, 128–52

Dosi, Giovanni, Christopher Freeman, Richard Nelson, Gerald Silverberg and Luc Soete (eds) (1988), *Technical Change and Economic Theory*, London, Pinter

Ergas, Henry (1987), Does technology policy matter? in: B.R. Guile and H. Brooks (eds), *Technology and Global Industry: Companies and Nations in the World Economy*, Washington, DC, National Academy Press

Freeman, Christopher (1988), Japan: a new national system of innovation? in: Dosi et al. (eds), 330–48

Granberg, A. (1997), Mapping the cognitive and institutional structures of an evolving advanced-materials field: the case of powder technology, in B. Carlsson (ed.), *Technological Systems and Industrial Dynamics, Economics of Science, Technology and Innovation* volume 10, Dordrecht, Kluwer Academic Publishers, 169–200

Hagedoorn, J. and J. Schakenraad (1990), Strategic partnering and technological cooperation, in B. Dankbaar et al. (eds), *Perspectives in Industrial Organization*, 171–191, Dordrecht, Kluwer

Håkansson, H. (1987), *Industrial Technological Development: A Network Approach*, London, Croom Helm

Håkansson, H. (1989), *Corporate Technological Behaviour; Cooperation and Networks*, London, Routledge

Kreiner, K. and M. Schultz (1991), Crossing the institutional divide: networking in biotechnology, paper presented at the international conference on 'Transnational Business in Europe', Tilburg University, March 20–22

Limpens, I., B. Verspagen and E. Beelen (1992) Technology policy in eight European countries: a comparison, A study carried out for the Dutch Ministry of Economic Affairs, MERIT, University of Limburg

Lundvall, Bengt-Åke (1988), Innovation as an interactive process: from user–producer interaction to the national system of innovation, in: Dosi et al. (eds), 349–69

Lundgren, Anders (1995), *Technological Innovation and Network Evolution*, Routledge, London

Mytelke, L. and M. Delapierre (1987), The alliance strategies of European firms in the information technology industry and the role of ESPRIT, *Journal of Common Market Studies* 26, 2, 231–53

Nelson, R.R. (ed.) (1993), *National Innovation Systems; A Comparative Analysis*, New York/Oxford, Oxford University Press

Nelson, R.R. and Rosenberg, N. (1994), Technical innovation and national systems, in R.R.Nelson (ed.), *National Innovation Systems; A Comparative Analysis*, 3–21

Niosi, J. (1994), Introduction: technical and organizational change in Western enterprises and the policy environment, in J. Niosi (ed.), *New Technology Policy and Social Innovations in the Firm*, London, Pinter, 1–15

Palda, Kristian S. (1986), Technological intensity: concept and measurement, *Research Policy* 15, 187–98

Peterson, J. (1993), Assessing the performance of European collaborative R&D policy: the case of Eureka, *Research Policy* 22, 243–64

Porter, M.E. (1990), *Competitiveness of Nations*, New York, The Free Press

Reger, G. and Kuhlmann, S. (1995), *European Technology Policy in Germany: The Impact of European Community Policies upon Science and Technology in Germany*, Heidelberg, Physica-Verlag

Teece, D.J. (1986), Profiting from technological innovation, *Research Policy* 15, 6, 78–98

Index